INTRODUCTION TO BIOTECHNOLOGY

INTRODUCTION TO
BIOTECHNOLOGY

For B.Tech.–Ist Year

By

DR. B.L. SAINI
Lecturer in Botany
University College
Kurukshetra University, Kurukshetra

UNIVERSITY SCIENCE PRESS
(An Imprint of Laxmi Publications Pvt. Ltd.)

BANGALORE • CHENNAI • COCHIN • GUWAHATI • HYDERABAD
JALANDHAR • KOLKATA • LUCKNOW • MUMBAI • RANCHI
NEW DELHI

Published by :
UNIVERSITY SCIENCE PRESS
(An Imprint of Laxmi Publications Pvt. Ltd.)
113, Golden House, Daryaganj,
New Delhi-110002

Phone : 011-43 53 25 00
Fax : 011-43 53 25 28

www.laxmipublications.com
info@laxmipublications.com

First Edition : 2010

OFFICES

✆ **Bangalore**	080-26 61 15 61	✆ **Chennai**	044-24 34 47 26
✆ **Cochin**	0484-237 70 04, 405 13 03	✆ **Guwahati**	0361-254 36 69, 251 38 81
✆ **Hyderabad**	040-24 65 23 33	✆ **Jalandhar**	0181-222 12 72
✆ **Kolkata**	033-22 27 43 84	✆ **Lucknow**	0522-220 95 78
✆ **Mumbai**	022-24 91 54 15, 24 92 78 69	✆ **Ranchi**	0651-221 47 64

UBT–9509–120–INTRO TO BIOTECHNOLOGY–SAI **C—1038/010/02**

Typeset at : ABRO Enterprises, Delhi. Printed at : Repro Knowledgecast Limited, Thane

Preface

The present book, Introduction to Biotechnology, is written strictly according to the syllabus of B.Tech.-Ist Year of Kurukshetra University, Kurukshetra. This book is of immense help to the students pursuing undergraduate courses in Biological Sciences in general and those who are studying B.Tech. Biotechnology or B.Sc. Biotechnology in particular as in it diverse basic concepts of Life science and Biotechnology have been explained in simple, easy to understand language. This book is divided into four units and contains a total of 12 chapters. The first unit of this book deals with the fundamental concepts of matter, life, and energy. It has been a topic of concern to the man of all ages to know how and when life originated. Is it present only on the earth or also elsewhere ? This topic is discussed in the beginning of the first unit. The other topics included in this unit are: cell as the basic structural and functional unit of life; structure of prokaryotic and eukaryotic cell; biomolecules and enzymes. The second unit of the book is related to diversity of life systems. In this section some important and basic aspects of plant system, animal system and microbial system are discussed. These three types of systems are necessary for the sustenance of the life on the planet earth. The third unit presents an insight about the basic and fundamental concepts of organic evolution, Mendel's laws of inheritance, cell division, experimental evidences in support of nucleic acids (DNA or RNA) as the genetic material, and an introduction to current topics like genetic engineering, genomics and bioinformatics. The fourth unit is devoted to Biotechnology—the revolutionary science of the 21st century. Biotechnology has lots of scope and it is playing a significant role in the field of agriculture, industry, medicine and health care and in reducing the pollution and in the management of the environment.

Every concept explained in this book is well documented and supported by self explanatory neat and well labeled diagrams with latest information. During the course of writing of this book I received whole hearted support from my friends and fellow colleagues I acknowledge thanks to all of them. I am very thankful to my family members, my wife Monica and my sons Ishan and Dipank for sparing me from the household chores while writing the book. My younger son Dipank who is in 7th standard deserves special thanks as most of the diagrams in the present book are drawn with his help.

—Author

Syllabus

Ist YEAR B.TECH
INTRODUCTION TO BIOTECHNOLOGY
(BT-101E)

L T P/D
3 1

Theory : 100 marks
Sessional : 50 marks
Total : 150 marks
Time : 3 hrs.

UNIT–I

Introduction to Life : Characteristics of living organisms. Hierarchy of organisation and factors responsible for regulating different levels of organisations. Structure of prokaryotic and eukaryotic cell. Basic concept of state and homeostasis.

Introduction to Biomolecules : Definition, general classification and important functions of carbohydrates, lipids, proteins, nucleic acids and vitamins.

Enzymes as Biocatalysts : General characteristics, nomenclature and classification of enzymes. Effect of temperature, pH, enzyme and substance concentrations on the activity of enzymes. Elementary concept of cofactors and coenzymes.

UNIT–II

Plant System : Basic concepts of plant growth, nutrition, photosynthesis and nitrogen fixation. Types of growth regulators and their physiological effects.

Animal System : Elementary study of digestive, respiratory, circulatory, excretory systems and their functions.

Microbial System : History of microbiology, Types of microbes and properties, Economic importance and control of microbes.

UNIT–III

Evolution : Theories of evolution, Mendel's laws of inheritance, variation and speciation.

Genetics : Cell division—mitosis and meiosis. Evidence of nucleic acids as a genetic material, Central dogma.

Genetic Engineering : Elementary knowledge of recombinant DNA technology, bioinformatics and genomics.

UNIT–IV

Introduction to Biotechnology : Definition, scope and achievements. Tools used in biotechnology.

Applications of Biotechnology in Agriculture, Medicine and Environment—An elementary knowledge.

Prospects and Public Perception of Biotechnology.

Note : The Examiners will set eight questions, taking two from each unit. The students are required to attempt five questions in all selecting at least one from each unit. All questions will carry equal marks.

Contents

UNIT—1
(1—50)

Chapter 1 : Introduction to Life **3—16**

Characteristics of Living Organisms 4
Hierarchy of Organisation 6
Cell and Cell Theory 7
Prokaryotic and Eukaryotic Cells 8
Eukaryotic Cell 10
Steady State 14
Homeostasis 15
Exercises 16

Chapter 2 : Biomolecules **17—39**

Carbohydrates 17
Lipids 20
Proteins 23
Nucleic Acids 29
Nucleoside and Nucleotide 31
Vitamins 36
Exercises 38

Chapter 3 : Enzymes **40—49**

Characteristics of Enzymes 40
Mode of Enzyme Action 42
Mechanism of a Biochemical Reaction 43
Nomenclature and Classification of Enzymes 44
Factors Affecting Enzyme Activity 45
Inhibition of Enzyme Activity 47
Coenzymes 48
Cofactors 48
Exercises 49

UNIT—2
(51—110)

CHAPTER 4 : Plant System 53—73

Cryptogams 53
Phanerogams 53
Parts of an Angiospermic Plant 54
Mineral Nutrition in Plants 54
Nitrogen Metabolism in Plants 59
Photosynthesis 61
Plant Growth 70
Exercises 73

CHAPTER 5 : Animal System 74—95

Nutrition in Animals 75
Human Digestive System 76
Human Respiratory System 81
Mechanism of Respiration 83
Circulation of Body Fluid in Animals 84
Human Circulatory System 86
Human Excretory System 93
Exercises 95

CHAPTER 6 : Microbial System 96—110

History of Microbiology 96
Types of Microbes and their Properties 99
Economic Importance of Microorganisms 103
Harmful Activities 107
Control of Microbes 108
Exercises 110

UNIT—3
(111—154)

CHAPTER 7 : Organic Evolution 113—124

Religious View (Theory of Special Creation) 113
Scientific View (Doctrine of Evolution) 113
Evidences in Support of Evolution 114
Theories of Evolution 118
Variation and Speciation 121

Speciation 123
Exercises 124

Chapter 8 : Mendel's Laws of Inheritance **125—133**
Terminology Used in Genetics 126
Exercises 133

Chapter 9 : Cell Division **134—142**
Cell Cycle 134
Mitosis 135
Meiosis 137
Exercises 142

Chapter 10 : Nucleic Acids : The Genetic Material **143—153**
Evidences in Support of that DNA is the Genetic Matter 144
Recombinant DNA Technology 149
Gene Library 151
Genomics 152
Exercises 153

UNIT—4
(155—182)

Chapter 11 : Biotechnology : Scope, Achievements and Techniques **157—169**
Definition of Biotechnology 158
Scope and Achievements of Biotechnology 158
Tools and Techniques of Biotechnology 159
Transgenic Plants 164
Methods of Gene Transfer in Plants 164
Cloning, Clones and Biotechnology 166
Exercises 169

Chapter 12 : Applications of Biotechnology **170—182**
Application in Agriculture 170
Application in Medicine 174
Application in Environment 179
Biotechnology and its Public Perception 181
Exercises 182

Index 183—186

UNIT 1

- ★ Introduction to Life
- ★ Biomolecules
- ★ Enzymes

Chapter 1 INTRODUCTION TO LIFE

In this universe life exists only on the planet Earth as far as knowledge is available with us. Universe is extremely vast and has no beginning or ending and is believed to be continuously expanding. In this never ending, ever expanding and unlimited universe which is an aggregation of all matter, energy and space living organisms inhabited only on the earth of the solar system. Solar system is a part of the Milky Way galaxy. A galaxy is a huge assemblage of stars, dust and gases. There are billions of galaxies in the universe. Scientists are exploring the possibilities of existence of life elsewhere in the universe but till now no traces of life have been reported from any other part of the universe. So far creating of life by artificial means is a distant and unfulfilled dream of scientists. Various objects living and non-living present in the world basically composed of matter. Matter means something which has mass and occupies space. Organization of matter differs in living organisms and non-living objects. In living organisms matter is organized in specific manner where as in non-living objects matter is present in an unorganized and random manner. The matter in the living organisms is integrated in such a manner exhibiting special properties which are not found in their random mixture.

A question that has puzzled the philosophers, thinkers, saints, scientists and layman of all the ages alike is to know how and when life originated. There are various theological speculations about the origin of life and almost every religion has its own explanation about the meaning and origin of life. These theological speculations about the origin and meaning of life though are of mythological significance but they lack a scientific basis and fail to explain the origin of life in a logical manner. According to the modern theory of origin of life, it originated by chemogenic process from the chemical compounds of the earth's atmosphere. The age of the earth is about 5.2 billion years which was formed as a small detached part of the sun. Initially it was a molten mass of hot gases which took thousands of years to cool down. Life originated on the earth about 3.9 billion years ago, by chemogenic method *i.e.,* by the aggregation of matter. During that time simple molecules like methane (CH_4), ammonia (NH_3), water vapours (H_2O) and hydrogen (H_2) present in the earth's atmosphere collided with the help of cosmic forces like UV-radiations, lightening etc., to generate sugars, fatty acids, amino acids, nucleotides etc., which on further condensation formed biomolecules like DNA, RNA, proteins, carbohydrates, lipids etc., and then by the aggregation of these molecules primitive life forms came into existence in an aquatic environment *i.e.,* the oceans. Early life forms were heterotrophic, anaerobic and unicellular organisms.

Living system is an open system *i.e.,* it exchanges both matter and energy from its surroundings. All living organisms with the exception of viruses have unanimity in their composition as they are made up of cell or cells. Cell is the structural and functional unit of life. It has a unique property of self replication. Cells are basically of two types : prokaryotic and eukaryotic. Bacteria and cyanobacteria possess prokaryotic cells and plants and animals possess eukaryotic cells. Plant cells possess chlorophyll, a green pigment which helps in conversion of solar energy into chemical energy. As plants are autotrophs and synthesizes their own food by converting carbon dioxide and water into sugars and releasing oxygen in the environment. Plants by virtue of this character make the life on the earth a self sustainable system as all other organisms obtain their food (energy) directly or indirectly from the plants. Oxygen is utilized by the living organisms for the oxidation of food to release energy and energy is required by each and every living organism to carry out their life functions.

Viruses do not have cellular body organization. A virus is an ultramicroscopic entity composed of a protein coat and either of the nucleic acid DNA or RNA. Viruses exhibit characters similar to living as well as non-living objects. When a virus is inside its host it shows characters similar to a living organism and at the same time it can be stored in a container similar to any other chemical compound.

CHARACTERISTICS OF LIVING ORGANISMS

All the vital life activities carried out by the living organisms is due to protoplasm which is present in all the living cells. **Huxley** in 1868, explained that *protoplasm is the physical basis of life*. Protoplasm of all the living organisms is fundamentally similar in its physico-chemical nature and biological properties. Yet at micro level, protoplasm of every organism is different. Living organism show lots of diversities yet they have some common characteristics.

1. Shape and Size

Every living organism unicellular or multicellular has a peculiar shape and size. One can recognize easily a cow, an elephant or a mango tree. Every organism pass through juvenile, young and adult stage before it dies.

2. Growth

Every living organism grows and growth is achieved either by increase in cell size or increase in cell number or both. It is an irreversible change in which increase in size of the organism accompanied by gain in weight take place.

3. Metabolism

All living organisms show metabolic activities which are of two types :

(*i*) ***Anabolism.*** It is a synthetic activity in which macromolecules are synthesized from simple molecules. For example, during the process of photosynthesis green plants synthesize carbohydrates (glucose) from simple compounds like carbon dioxide and water with the help of solar energy.

$$6CO_2 + 12H_2O \xrightarrow{\text{Chlorophyll}} \underset{\text{Glucose}}{C_6H_{12}O_6} + 6H_2O + 6O_2$$

(*ii*) ***Catabolism.*** In such activities, large molecules are broken into small and simple molecules in which energy is released. This energy is stored in the energy rich molecule—adenosine triphosphate (ATP) and is utilized by the organism when needed. Cellular respiration is a catabolic process in which a glucose molecule is oxidized with the help of enzymes into carbon dioxide and water to release energy.

$$\underset{\text{Glucose}}{C_6H_{12}O_6} + \underset{\text{Oxygen}}{6O_2} \longrightarrow \underset{\text{Carbon dioxide}}{6CO_2} + \underset{\text{Water}}{6H_2O} + \text{Energy (674 kcal)}$$

4. Adaptation

It means adjustment by an organism to its surroundings. Living organisms adapt to changes in their environment for their better survival. There are two types of adaptations : Short-term adaptation and Long-term adaptation.

Short-term adaptation is during the life span of an organism such as, migration of birds. Each year, Siberian Cranes migrate from Siberia to marshy lands of the plains of North India during winter season for food, breeding areas and other better environmental conditions and they go back from India to Siberia in the summer season. Similarly, Invertebrates hibernate and aestivate for their survival. In winter season, Frog go deep into mud and remain there till the winter is over. This is also known as winter sleep, during this period, there is negligible metabolic activity by it. *Pila* performs summer sleep (aestivation). In summer, there is scarcity of water so it closes its mouth by a lid and withdraws itself into a hard calcareous body shell and remain buried in mud till the rainy season.

In Long-term adaptation, changes in the genetic material which is inheritable and is achieved for the survival of living organisms for example, xerophytes growing in deserts have modified in number of ways to face harsh environment conditions. These plants have deep roots system, narrow leaves, thick cuticle, and sunken stomata for the conservation of water.

5. Respond to Environment

Plants, animals and microorganism react to different degree when some change in their environment take place. If we touch the leaves of *Mimosa pudica* (touch me not plant) they immediately droop down. If animals are exposed to high temperature they try to move away from it and reach at a place of optimal temperature. Similarly if *Amoeba* is exposed to high light intensity it moves to low intensity area.

6. Excretion

During metabolic processes, some by-products are formed which are of no use to the organism and these are to be eliminated from the body otherwise they may become toxic to the organism *e.g.*, during breakdown of proteins in human body, ammonia is formed. Excess of ammonia is toxic to our body so it is not allowed to accumulate. It is eliminated in the form of urea and uric acid in urine.

7. Reproduction

Every living organism creates new organisms similar to themselves. This is necessary for the perpetuation of an organism. It is of two main types : Asexual and Sexual. In asexual reproduction a single parent produces progeny of its own kind which is a clone *i.e.*, it is genetically similar to its single parent. It occurs in some invertebrates and lower plants.

In sexual reproduction there is a requirement of male and female gametes from adult male and female of the same species. These two gametes fuse to form zygote from which a new individual is born which shares its characters from both of its parents. So sexual reproduction creates variations and also helps in evolution.

8. Repair and Regeneration

Organisms have a unique property of healing and repair. This property is more pronounced in lower organisms as compared to higher organisms. Some lower animals like *Planaria* has highest degree of regeneration. An adult *Planaria* if cut into two or more pieces, each piece reorganizes into a new *Planaria*. Similarly, if tail of a wall lizard is bitten by its enemies it is regenerated after sometime. Likewise a cut made on a mango tree is healed up in the due course of time.

9. Death

Every organism after completing its expected life span dies. Death means cessation of all life activities and it is an important happening for the continuity of the life. The different mineral elements required by living organisms are present in fixed amount in the earth crust. These elements are taken from the earth crust by the organisms and get accumulated in their bodies in a locked form. In this way their amount goes on depleting from the earth crust. If it is allowed continuously, no mineral will be left for the young ones. So when organisms die, their body is decomposed by saprophytic bacteria and fungi and the minerals get released and are again added in the soil for the availability of other organisms.

HIERARCHY OF ORGANISATION

As shown in Fig. 1.1, atom is the basic unit of organization in both living organisms and non-living objects. Atom is sub microscopic and the smallest unit of an element. There are about 100 elements, like Carbon (C), Hydrogen (H), Nitrogen (N), Oxygen (O), Phosphorus (P), Sulphur (S), Potassium (K), Calcium (Ca) etc. Two or more similar or dissimilar atoms unite to form a molecule. Some molecules are small *e.g.*, Oxygen(O_2), Carbon dioxide (CO_2), Water (H_2O), Ammonia (NH_3),Methane (CH_4) etc., others are large called macromolecules. In a macromolecule small monomer molecules get polymerized. Important macro biomolecules are nucleic acids, proteins, lipids, carbohydrates, vitamins etc. Various micro and macromolecules get aggregated in a characteristic manner to form a cell. A **cell** is the simplest self-duplicating unit and it obtains energy from sun (autotrophs) or from food (heterotrophs) and uses this energy for growth and reproduction. The exact evolution and development of cellular life out of aggregation of biomolecules has yet to be explained fully.

The next higher level of organization is **tissue.** Tissue is a group of similar cells performing a specific function. There are four main types of tissue in animals : epithelial, muscular, connective and nervous tissue. A group of tissues constitutes an **organ**. Stomach is an organ of digested system and has all the four types of tissues. A group of organs collectively form an **organ system** *e.g.*, organs like salivary glands, stomach, intestine, liver, pancreas, etc., are all part of the digestive system which helps in digestion of food. Similar to digestive system there are other organ systems like respiratory system, blood vascular system, excretory system, nervous system etc., which all are parts of an **organism.** An organism which is similar in morphological characters and can inter breed is a species. A group of members of the same species residing in a geographical area is known as **population.** Population of different types of organisms present in an area constitutes **community**. Different organisms like plants, animals and microorganisms of a community interact with its physical environment comprising of soil, air, light, water, etc., forms an **ecosystem.** There are various types of ecosystem like pond, grassland, desert, forest, etc. The highest level of living organisms is **biosphere**. It consists of lithosphere (upper most crust of the earth), hydrosphere (oceans and other water bodies) and atmosphere near the earth surface where life exists.

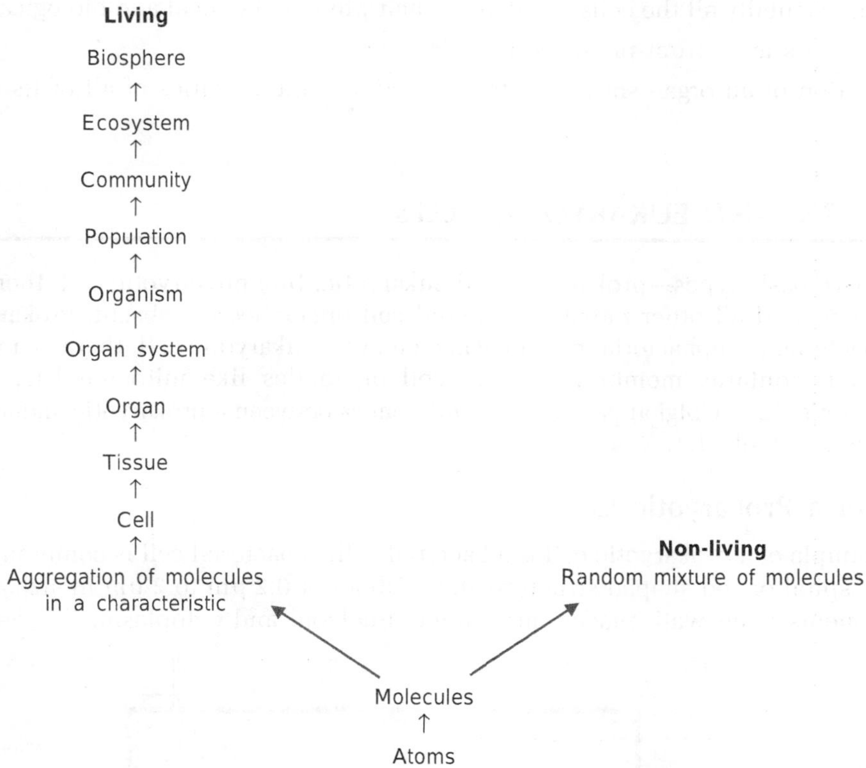

Fig. 1.1. Hierarchy of organization of life.

CELL AND CELL THEORY

Cell is the structural and functional unit of all living organisms except viruses. Viruses are sub-cellular living molecules composed of a protein coat and either type of nucleic acid—DNA or RNA. They behave as living organisms only inside their host. Once outside the host they are like any other chemical compound. The term cell was coined by **Robert Hooke** in 1665. He cut a thin section of cork and observed it under his microscope and he found a honeycomb like pattern in it. He called these small compartments as cells. Actually, Robert Hooke saw the empty dead cells with their walls. But owing its historical importance the term cell has been retained. **Robert Brown** while studying the orchid roots in 1831, found a prominent round structure in it and named it nucleus. Nucleus controls the activity of a cell and it is found in all living cells.

Schleiden (1838), studied a number of plants and found that their body is composed of cells. At the same time **Schwann** studied a number of animals and reached to the same conclusion. Both of them proposed jointly cell theory. They did not explain how new cells originate. This was explained later by **Virchow** in 1855 that new cells arise from pre-existing cells. Salient points of the cell theory as proposed by **Schleiden** and **Schwann** with some modifications are as follows :

1. Body of all the living organisms is composed of cells and their products.
2. Fundamentally all the cells are alike in their physicochemical and biological properties.
3. New cells arise from pre-existing cells.
4. Function of an organism is due to interactions and activities of all of its constituents cells.

PROKARYOTIC AND EUKARYOTIC CELLS

Cells are of two basic types—prokaryotic and eukaryotic. In a prokaryotic cell, there is no well defined nucleus and all other membrane bound cell organelles are absent. Prokaryotic cell is present in bacteria, cyanobacteria, mycoplasma, etc. In a eukaryotic cell, there is a well defined nucleus and it contains membrane bound cell organelles like mitochondria, chloroplast, endoplasmic reticulum, Golgi apparatus, etc. Differences between a prokaryotic and an eukaryotic cell are shown in table 1.1.

Structure of a Prokaryotic Cell

A typical example of a prokaryotic cell is a bacterial cell. A bacterial cell is commonly spherical, comma like, spiral or rod shaped structure and is about of 0.2 μm to 2 μm in size and has four main components : cell wall, plasma membrane, nucleoid and cytoplasm.

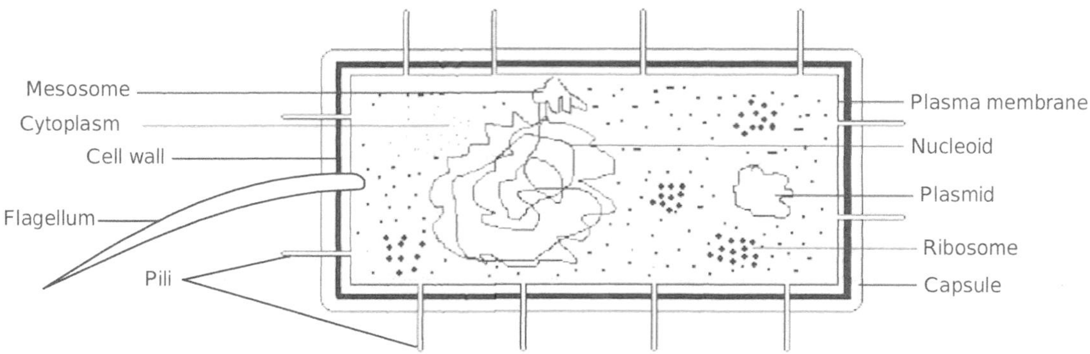

Fig. 1.2. Structure of a Bacterial cell.

1. Cell Wall

It is the outermost part of the bacterial cell and is comprises mainly of peptidoglycan (murein or mucopeptide). Peptidoglycan is unique to the cell wall of prokaryotes. It is composed of long chains of alternating units of N-acetyl glucosamine (NAG) and N-acetyl muramic acid (NAM). There are many such parallel chains which are joined by small peptides of a few amino acids. Besides; it there are some traces of lipo proteins, lipo polysaccharide and teichoic acid. Cell wall is tough and rigid. It provides shape to the cell and also protects it. On the basis of composition of cell wall, bacteria are grouped into two groups : Gram-positive bacteria and Gram-negative bacteria.

In gram-positive bacteria, cell wall is single layered and smooth. It is about 20–80 nm thick. It is composed of 70 to 80 percent murein and 2 to 4 percent lipid. Besides, it also contains teichoic acid. A few pathogenic bacteria belong to gram-positive group.

In gram-negative bacteria, cell wall is two layered and wavy that comes in contact with plasma membrane only at a few loci. It is about 8–12 nm thick. It is composed of 10–20 percent murein, 20–30 percent lipids. Teichoic acid is absent in the cell wall of gram-negative bacteria. Most of the pathogenic bacteria belong to gram-negative group.

In some pathogenic bacteria, there is an additional gelatinous layer outer to the cell wall and is known as **capsule.** A capsule is composed of polysaccharides and is sticky in nature due to which a pathogenic bacterium sticks with its host. Capsule also protects the bacteria from desiccation and antibodies.

2. Cell Membrane

Inner to bacterial cell wall there is a thin membrane surrounding the cytoplasm known as cell membrane or plasma membrane. It is composed of protein and phospholipids. Respiratory enzymes are also present on this membrane. It is a selectively permeable membrane which regulates the movement of nutrients in the cell. Plasma membrane is associated with structures like mesosome, flagella and pili.

(*i*) ***Mesosome.*** It is a fold of plasma membrane aggregated at a place and increases the surface area of the membrane and is rich in enzymes. It is also the place where nucleoid is attached with it. A septum formation is initiated at this point during the binary fission.

(*ii*) ***Flagella.*** A bacterial cell may be associated with one or more flagella. A flagellum is a hair like process arises from the basal granule present in the cytoplasm near the membrane and come out from the cell in the form of a long fibre. It is made up of protein flagellin. In a flagellum, there are eight parallel strands of flagellin. Flagella help in locomotion.

(*iii*) ***Pili.*** Some of the bacterial cells are associated with fine hair like processes of about 0.5 to 2 μm in length which arises from the plasma membrane and are present on the cell surface.These are made up of pilin protein. Pili help in attachment of a bacterial cell on the substratum. Some pili also act as conjugation canal through which DNA of a bacterial cell passes into another bacterial cell. Pili are present generally in gram-negative bacteria.

3. Nucleoid

Bacteria do not have a well defined nucleus. The genetic matter of bacterium is a highly coiled and circular DNA molecule of about 1000 μm long. This bacterial chromosome is present in the cytoplasm and is generally attached with the mesosome.

4. Cytoplasm

It is a jelly like fluid composed of aqueous solution of inorganic and organic substances. Cytoplasm is associated with the following components :

(*i*) ***Ribosomes.*** These are minute spherical particles scattered in the cytoplasm. These are composed of RNA and Protein. Ribosomes help in protein synthesis. Ribosomes of prokaryote are of 70S type.

(*ii*) ***Plasmid.*** A bacterial cell may possess one or more plasmid. A plasmid is an extra chromosomal DNA, contains a few genes. Three important genes associated with the plasmid are F factor, R factor and Col factor. F factor is also known as fertility factor and helps in conjugation in bacteria. R factor is also known as resistance factor and it provides immunity against antibiotics to the bacteria. Col factor produces colchicine which kills sensitive *E. coli* cell.

(*iii*) ***Other Inclusions.*** There are various type of cell inclusions present in the cytoplasm and include reserve food (lipids, protein and glycogen), volutin granules containing reserve phosphates and small quantity of basic proteins and RNA. Some gas vacuoles are also present in the cytoplasm.

In photosynthetic bacteria, there are some lamellae present in the cytoplasm which contain bacteriochlorophyll to carry out the process of bacterial photosynthesis.

EUKARYOTIC CELL

A eukaryotic cell has a well defined nucleus and is bounded by a nuclear membrane. Plant cell and animal cell are examples of eukaryotic cells. Eukaryotic cells are of variable shape and size. They may be spherical, ovoid, cylindrical or polygonal in shape and about 10 to 20 μm in size.

Plant Cell

A plant cell is composed of cell wall, plasma membrane, cytoplasm and nucleus. Structure of a plant cell is shown in Fig. 1.3.

1. Cell Wall

It is the outermost part of the plant cell. It is rigid, dead in nature and is permeable to both solutes and solvents. It is mainly composed of cellulose, hemicelluloses. Adjoining plants cell are joined by a cementing substance known as pectin. In between walls of two plants cell there are some narrow channels called plasmodesmata. Plasmodesmata facilitates intercellular transport of substances. Cell wall protects the cell and also gives shape to it.

2. Plasma Membrane

It is the limiting membrane of the cell inner to the cell wall and surrounds the cytoplasm. It is made up of lipoprotein which has almost equal amount of protein and phospholipids. This is a selectively permeable membrane. It allows the movement of some of the substances across it, not all. For this purpose, there are certain tunnel proteins and carrier molecules in it.

3. Cytoplasm

Inside the plasma membrane and outside the nucleus there is jelly like viscous fluid known as cytoplasm. It is an aqueous solution of inorganic and organic compounds and contains different cell organelles. A cell organelle is a compartment of the cell which performs some specific function. Different cell organelles present in the cytoplasm are as follows :

(*i*) ***Mitochondria.*** Dispersed throughout the cytoplasm, inside the cells are present many small sausage shaped structures called mitochondria. These are the seats of respiration and are known as **powerhouse** of the cell. Mitochondrion contains enzymes of Krebs cycle. Oxidative phosphorylation also takes place in them. The energy released during this process is stored in the form of ATP molecules.

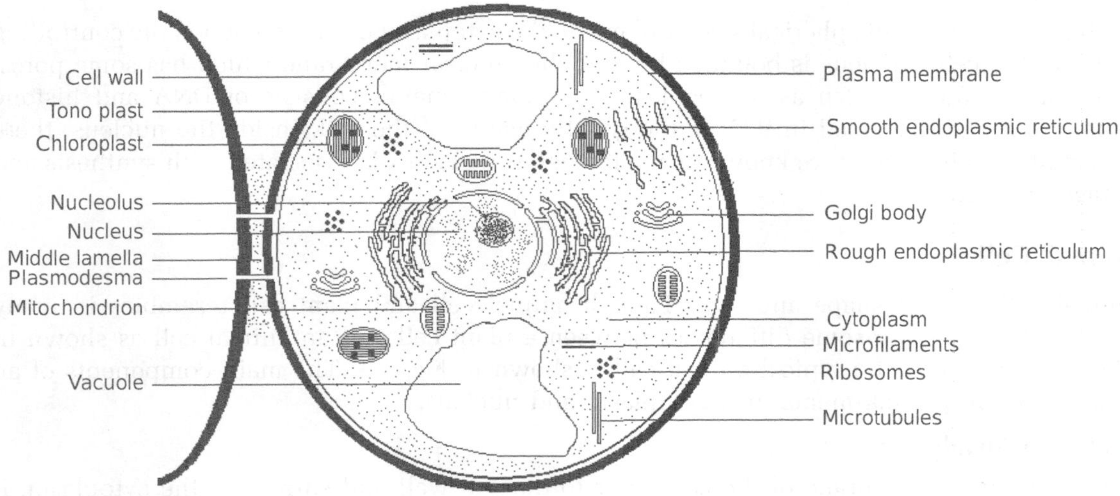

Fig. 1.3. Structure of a Plant cell.

(*ii*) ***Chloroplast.*** These are small, oval shaped structures found only in the plant cells. It contains photosynthetic pigments—chlorophylls and carotenoids in their lamellae. It performs the function of photosynthesis. In this process, solar energy is converted into chemical energy.

(*iii*) ***Endoplasmic Reticulum.*** It consists of narrow and interconnecting channels. It is of two types—Rough Endoplasmic Reticulum (RER) which is associated with ribosomes and Smooth Endoplasmic Reticulum (SER). Endoplasmic reticulum is associated with the synthesis and intracellular transport of proteins, lipids and carbohydrates.

(*iv*) ***Golgi Complex.*** It is also known as dictyosomes or Golgi body. A Golgi complex consists of stacks of small narrow channels called cisternae. It assists in synthesis and packaging of various types of secretions.

(*v*) ***Vacuoles.*** They are generally found in plant cells. In a mature plant cell, a vacuole is of very big size which covers major portion of its lumen. Each vacuole is surrounded by a membrane called tonoplast. Vacuole is a misnomer as it is not an empty space and contains a fluid called cell sap. It regulates the osmotic flow of substances in and out of the cell.

(*vi*) ***Microbodies.*** Peroxisomes and glyoxisomes are small membranous sacks found in plant cells. Peroxisomes are also associated with the glycolate metabolism and photorespiration in C3 plants. Glyoxisomes are present in oil bearing seed plants and are associated with conversion of oils into carbohydrates.

(*vii*) ***Ribosomes.*** Scattered throughout the cytoplasm, minute structures called ribosomes. Ribosomes of eukaryotic plant cell are of 80S type. They are composed of RNA and protein and help in the process of biosynthesis of protein.

(*viii*) ***Microfilaments and Microtubules.*** These are very fine structures and are made up of proteins actin and tubulin respectively and are distributed throughout the cytoplasm. They act as cytoskeletal structures and help in withholding of different cell organelles in the cytoplasm.

(*ix*) ***Cilia and Flagella.*** Cilia and flagella are structures associated with the locomotion present on the cell surface. In plants they are present only in zoospores and gametes of cryptogams.

4. Nucleus

Nucleus is a prominent spherical structure inside the cytoplasm of the cell and is the controlling center of the cell. Nucleus is bounded by a double nuclear membrane which has some pores. It contains a fluid known as nucleoplasm. Chromatin matter consists of DNA and histone protein is found dispersed in it. It is the genetic matter of the cell. Inside the nucleus, there is a small roundish structure known as nucleolus. A nucleolus is associated with synthesis and storage of RNA.

Animal Cell

Animal cell vary in shape and size. Though animal cell and plant cell resembles in many respect, still there are some differences between a plant cell and an animal cell as shown in table 1.2. Structure of a typical animal cell is shown in Fig. 1.4. The main components of an animal cell are plasma membrane, cytoplasm and nucleus.

1. Plasma Membrane

It is the limiting membrane of the cell inner to the cell wall and surrounds the cytoplasm. It is made up of lipoprotein which has almost equal amount of protein and phospholipids. This is a selectively permeable membrane. It allows the movement of some of the substances across it not all. For this purpose there are certain tunnel proteins and carrier molecules in it.

2. Cytoplasm

Similar to a plant cell, inside the plasma membrane and outside the nucleus there is jelly like viscous fluid known as cytoplasm. Different cell organelles present in the cytoplasm are as follows :

(*i*) ***Mitochondria.*** The description of mitochondrion is similar as in plant cell already described.

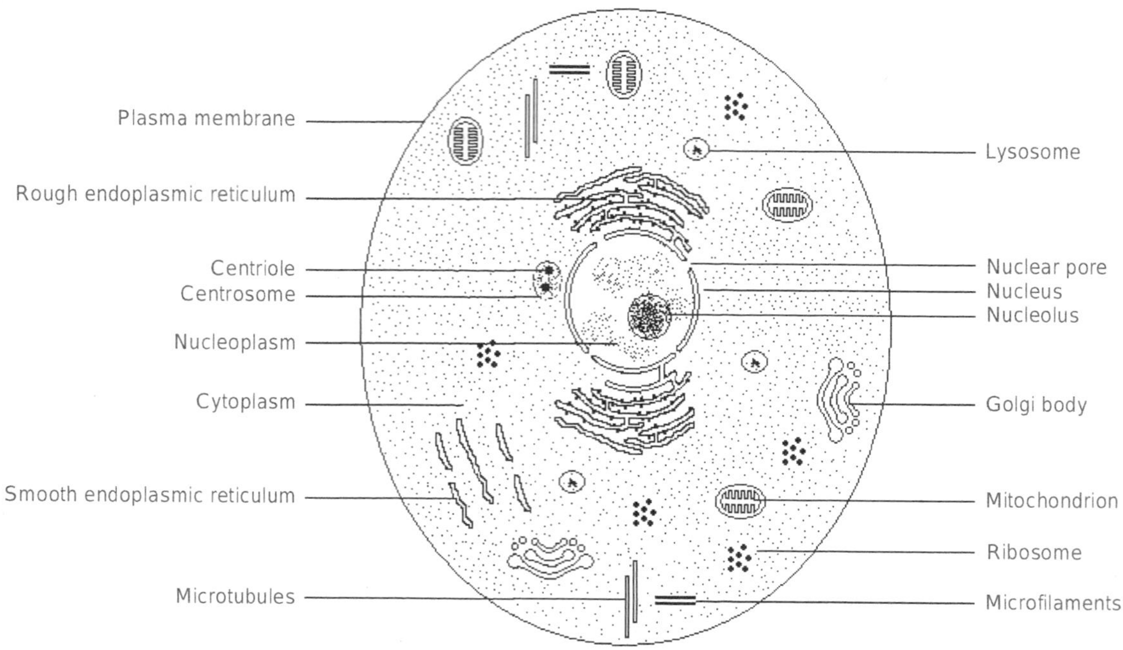

Fig. 1.4. Structure of an Animal cell.

(*ii*) ***Endoplasmic Reticulum.*** It has already been described in a plant cell and is applicable to animal cell also.

(*iii*) ***Golgi Complex.*** There are fewer number of golgi bodies in an animal cell as compared to a plant cell. Its structure and function are similar as described in a plant cell.

(*iv*) ***Lysosomes.*** It is commonly found in animal cell. It is a small membranous structure and contains about forty different types of acid hydrolase enzymes. It is involved in intracellular digestion of foreign matter. It is also known as *suicidal bag* of the cell. If it gets ruptured, it can digest the components of the cell in which it is present. During metamorphosis in frog the tadpole is converted into a young frog. In this process, the tail of the tadpole is digested by the enzymes of the lysosomes.

(*v*) ***Centrosomes.*** It is only present in animal cell and found near the nucleus. It consists of two dark staining bodies—centriole and centrosphere (cytoplasm around the centriole). It helps in the process of cell division.

(*vi*) ***Ribosomes.*** Its description is similar to a plant cell.

(*vii*) ***Microfilaments and Microtubules.*** Composition of microfilament and microtubules and their function is similar to a plant cell.

(*viii*) ***Cilia and Flagella.*** They are commonly found in animal cell. They are present both in cell of lower animals (*Paramecium, Giardia*) and higher animals like ciliated epithelium of human. Cilia and flagella are fine hair like structures present on the cell surface. Flagella are longer and cilia are smaller and have characteristic 9+2 arrangement. There are nine peripheral duplex and two central singlet strands of microtubules in flagellum/cilium. Each flagellum/cilium arise from a basal granule present in cytoplasm near the plasma membrane and are made up of protein flagellin.

3. Nucleus

An animal cell contains a centrally placed nucleus. The description of the nucleus is same as describe in the structure of a plant cell.

Table 1.1. Differences between eukaryotic cell and prokaryotic cell

S. No.	*Eukaryotic Cell*	*Prokaryotic Cell*
1.	A eukaryotic cell possesses a well define nucleus *i.e.,* nucleus has a distinct nuclear membrane enclosing nucleoplasm, containing variable number of chromosomes in it.	A prokaryotic cell does not possesses well define nucleus. Instead it is called nucleoid which consist a single circular chromosome lying freely in the cytoplasm or may be attached with the mesosome.
2.	Its chromosome consists of DNA and histones (nucleoprotein).	Its chromosome consists of only DNA. It is not associated with histones.
3.	A eukaryotic cell has a number of membrane bound cell organelles like, mitochondria, chloroplast, ER, Golgi body, etc.	All membrane bound cell organelles are absent in a prokaryotic cell.
4.	Cell wall of eukaryotic cell, when present as in a plant cell composed of cellulose.	Cell wall of prokaryotic cell is composed of peptidoglycan.
5.	Ribosomes are of 80S types.	Ribosomes are of 70S types.
6.	Flagella of eukaryotic cell have a characteristic 9+2 organization composed of protein tubulin.	Flagella of prokaryotic cell are simple in organization composed of 3–8 coiled strands of protein flagellin.

7.	Photosynthetic lamellae containing chlorophylls of a eukaryotic plant cell is present in a well organized manner inside the membrane of chloroplast.	Photosynthetic lamellae containing chlorophyll if present as in cyanobacteria (blue green algae) and some autotrophic bacteria lie freely in the cytoplasm not covered by a membrane.
8.	Microtubules and microfilaments are present which act as a cytoskeleton structure in the cytoplasm.	It lacks microtubules and microfilaments.

Table 1.2. Differences between a plant cell and an animal cell

S. No.	*Plant Cell*	*Animal Cell*
1.	A plant cell has a cell wall made up of cellulose and it is present outside the cell membrane.	An animal cell lacks a cell wall. Plasma membrane is the outermost covering of an animal cell.
2.	Chloroplasts are present.	Chloroplasts are absent.
3.	Vacuoles are present.	Vacuoles are generally absent in animal cell and if present are of very small size.
4.	Centrosomes are absent.	Centrosomes containing centrioles are present in animal cell.
5.	Lysosomes are absent.	Lysosomes are present.
6.	Cilia and flagella are generally absent in plant cell except in the gametes of cryptogams.	Cilia and flagella are frequently present in animal cell.

STEADY STATE

Biological world is like a physical world as both are composed of matter and energy. Energy is the capacity to do work. According to first law of thermodynamics, energy cannot be created nor be destroyed. It can be transformed from one form into another. Any energy transformation process is not 100% perfect as some energy is lost in the form of heat energy. Total energy of this world is constant.

Any living organism can be treated as a system. System means any part of the real world under study and it can interact with its surroundings. Surroundings mean everything that is not a part of the system. A system and its surroundings are always separated by real or imaginary boundaries across which energy and matter may be exchanged. A living organism is like an open system. An open system can exchange or transfer matter and/or energy between system and surrounding.

Living organisms perform many activities. Each and every activity requires energy. The source of energy to perform any activity by living organisms is ATP molecules which are also as energy currency of cell. ATP molecules are synthesized in the process of cellular respiration. It is a spontaneous and continuous physiological process. In this process, oxidation of food molecules such as glucose takes place. This is a catabolic process in which energy stored in glucose molecules is utilized in the production of ATP molecules. During this interconvertion some energy is lost in the form of heat.

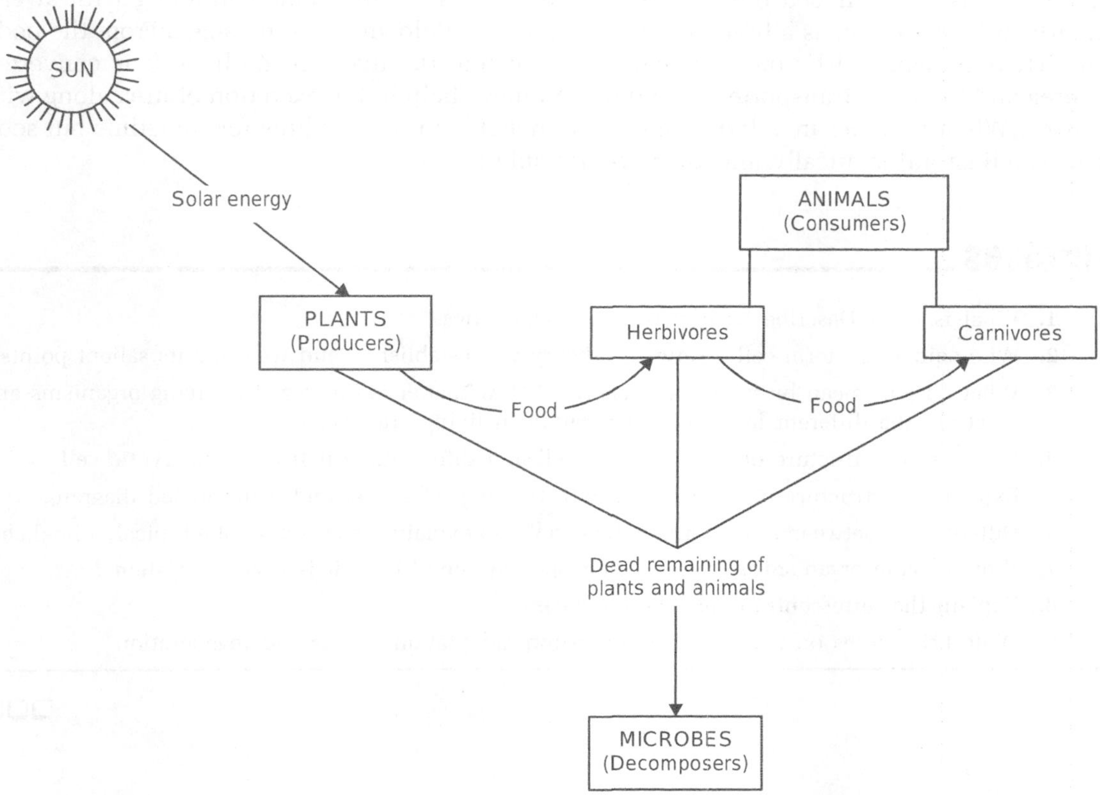

Fig. 1.5. Energy flow diagram showing sun as the ultimate source of energy to all the living organisms.

Respiration is a spontaneous process and according to second law of thermodynamics any spontaneous process leads to entropy. Entropy means randomness or disorder. If this process is allowed continuously then living system cannot be a viable system. But the loss of energy is made good by a living organism as it is an open system and there is an inflow of energy in it by way of intake of food in heterotrophs or directly from the sun in autotroph by the process of photosynthesis. Animals and most of the microbes are heterotrophs. Mainly animals are either herbivores taking food from a plant source or carnivores feeding on some other animals. Animals are called consumers. Microbes obtain their food from dead and decaying organic matter *i.e.,* dead remains of plants and animals. Sun is the ultimate source of energy to all the living organisms as shown in the energy flow diagram in Fig. 1.5.

HOMEOSTASIS

It means maintenance of constant condition of a cell or an organism in response to either internal or external changes. There are number of examples which can explain homeostasis, for example, regulation of body temperature. During a vigorous exercise, our body temperature rises but it is scaled down to normal temperature after sometime when we relax. Similarly, composition of our body fluid (blood and lymph) remains constant. During various catabolic processes, there is release of some waste materials. CO_2 is released during cellular respiration but it is not allowed to accumulate in the body fluid as it is expelled during expiration some amount of it is also diffused out by our skin in the form of carbonic acid along with sweat. Similarly, NH_3 is formed as a by-product during the breakdown of amino acids. From the body fluid NH_3 is transported by our circulatory system into the liver. In the liver, it is converted into urea and then it is transported to kidneys. Kidneys help in the excretion of urea along with the urine. When we enter in a dark room we are not able to see anything for sometime but soon our eyes adjust automatically and then we are able to see.

Exercises

1. What is life ? Describe its important characteristics.
2. Who coined the term cell ? How cell theory was established and what are its salient points ?
3. What do you mean by matter and energy ? How matter is organized in living organisms and what are the different levels of organization in living organisms ?
4. Describe the structure of a prokaryotic cell and differentiate it from a eukaryotic cell.
5. Explain the structure of a plant cell with the help of a neat and well labeled diagram.
6. Differentiate between a plant and animal cell and explain the structure of a typical animal cell.
7. Why a living organism is treated as an open system ? How it is a viable system ?
8. Explain the terms entropy and homeostasis.
9. Write brief notes on : metabolism, excretion, adaptation, repair and regeneration.

Chapter **2** # BIOMOLECULES

The organic molecules which are synthesized by the living cells only are known as **biomolecules.** Living cells have a unique property of converting simple inorganic nutrients taken from outside environment into complex organic molecules. These molecules act as either the structural components of cells or utilized to carry out various vital life functions. The important biomolecules are : carbohydrates, lipids, proteins, nucleic acids and vitamins.

CARBOHYDRATES

These are the compounds consisting of carbon, hydrogen and oxygen. The meaning of carbohydrates is hydrates of carbon as they have a general chemical formula $C_xH_{2x}O_x$. Carbohydrates are classified as monosaccharides, disaccharides and polysaccharides.

Monosaccharides

These are the simplest carbohydrates as they cannot be hydrolyzed further and have a general formula of $C_nH_{2n}O_n$. Monosaccharides are water soluble and sweet in taste and occur as trioses, tetroses, pentoses, hexoses and sedoheptuloses containing 3C, 4C, 5C, 6C and 7C atoms respectively. Monosaccharides may be aldoses or ketoses depending on whether they have an aldehyde or a keto group.

1. **Trioses.** These are the simplest sugars. Two important 3C sugars are glyceraldehyde and dihydroxyacetone and are formed as an intermediate compounds synthesized in the process of photosynthesis.

(*a*) (*b*)

Fig. 2.1. Triose Sugars (*a*) Glyceraldehyde (*b*) Dihydoxyacetone.

2. **Tetroses.** Erythrose $C_4H_8O_4$ is an important 4C sugar and is formed as an intermediate compound during dark reaction of photosynthesis.

3. **Pentoses.** Two important 5C sugars are ribose and deoxyribose. These are very important sugars as they are constituents of DNA and RNA. Ribose and deoxyribose sugars are constituents of nucleotides of RNA and DNA respectively.

(a) (b)

Fig. 2.2. Pentose sugars (*a*) Ribose (*b*) Deoxyribose.

4. **Hexoses.** Two important 6C sugars are glucose and fructose. Fructose is the sweetest naturally occurring sugar found in fruits. Glucose is an important sugar found in blood of animals. Glucose is also the chief respiratory substrates in all the organisms.

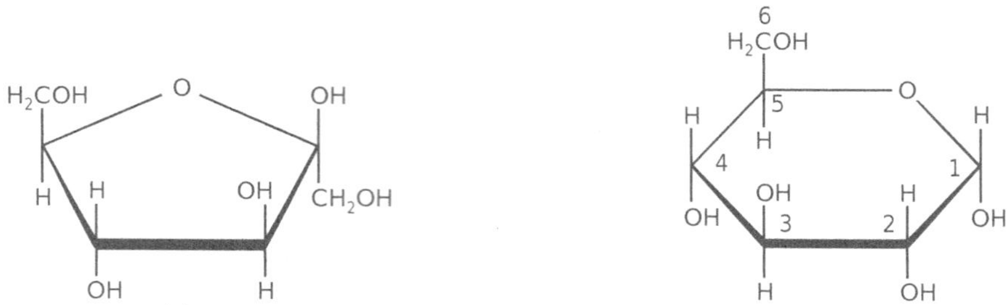

(a) (b)

Fig. 2.3. Hexose sugars (*a*) Fructose (*b*) Glucose.

5. **Sedoheptulose.** This is a 7C sugar and is formed as an intermediate compound during the dark reaction of photosynthesis and has a chemical formula $C_7H_{14}O_7$. All the monosaccharide aldoses or ketoses are reducing sugars as they reduce Fehlings solution and Benedicts solution.

Disaccharides

These are the sugars composed of two units of monosaccharide linked by a glycosidic linkage. Maltose, sucrose and lactose are three important disaccharides. All of these are water soluble and sweet in taste like monosaccharide.

1. **Maltose.** It is composed of a two units of glucose. It is also known as malt sugar and found in fruits. It is a reducing sugar.

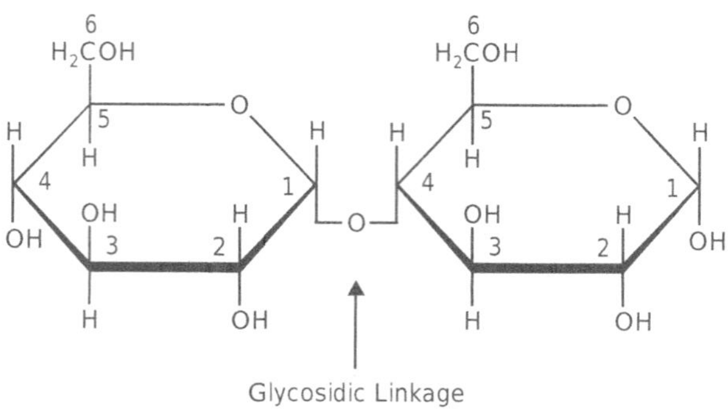

Fig. 2.4. Maltose sugar (a disaccharide).

2. **Sucrose.** It is composed of one unit each of glucose and fructose. It is also known as *cane sugar* and is obtained from sugar cane and beet. It is a non reducing sugar.
3. **Lactose.** It is also known as *milk sugar* as it is present in mammals. It is secreted in the mammary glands and is composed of one unit each of glucose and galactose. It is a reducing sugar.

Polysaccharides

These are complex carbohydrates consisting of a large number of monosaccharide. When a few number of monosaccharide are linked then it is known as oligosaccharide *e.g.,* Raffinose is a trisaccharide consists of one unit each of glucose, fructose and galactose. When hundred to thousands of monosaccharide are joined by glycosidic linkages then a polysaccharide is formed. Polysaccharides have high molecular weight. These are amorphous substances which are tasteless and are insoluble in water. Important polysaccharides are cellulose, starch and glycogen.

1. **Cellulose.** It is a linear polymer of β D-Glucose in which 300 to 2500 molecules of D-glucose are liked by β 1, 4-glycosidic linkage. It is a structural polysaccharide found in plant cell wall. Cellulose is the most abundant organic matter found on the earth.

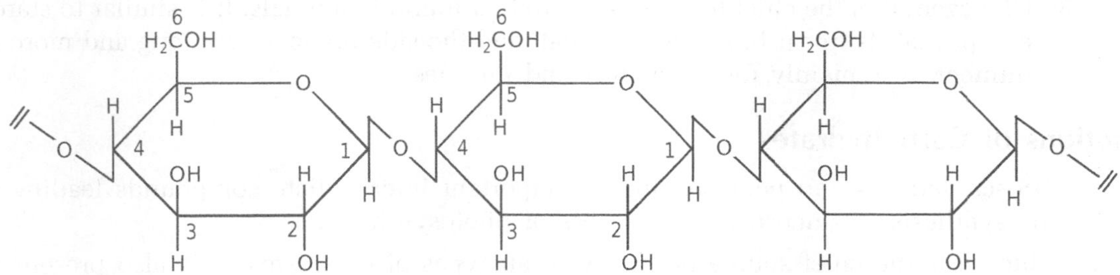

Fig. 2.5. Cellulose.

2. **Starch.** It is the chief food storage product found in plants. Starch is a polymer of α–D-glucose unit. It consists of two components—amylose and amylopectin. Amylose is a linear polymer consisting of hundreds of glucose monomers linked by α-1, 4 glycosidic linkages. Amylopectin is a brown chain polymer in it there are α-1, 6 glycosidic linkages in addition to the 1, 4 linkages.

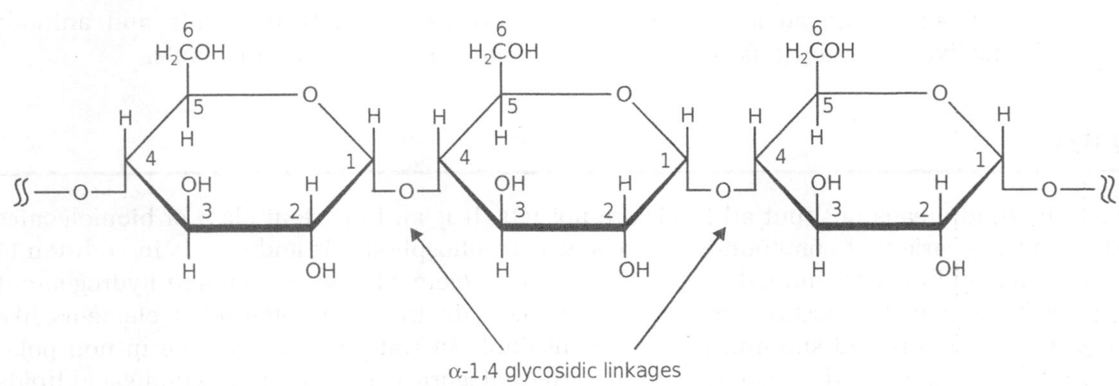

Fig. 2.6. Amylose (starch).

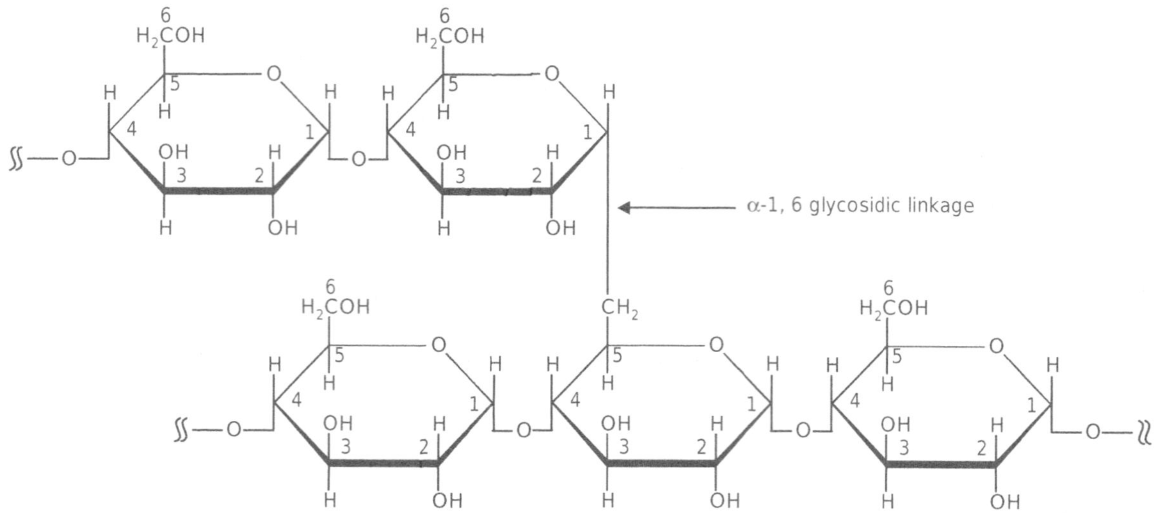

Fig. 2.7. Amylopectin (starch).

3. **Glycogen.** It is the chief food storage product found in animals. It is similar to starch except that its linear branches are small and the side branches are long and more in number. It is mainly found in liver and muscles.

Functions of Carbohydrates

1. Trioses, tetroses and pentoses are the important intermediate compounds leading to the synthesis of glucose in the process of photosynthesis.
2. Glucose is the chief source of energy in all types of organisms. It is also present as blood sugar in many animals.
3. Intermediate compounds like keto acids formed during the breakdown of glucose act as carbon skeleton for amino acids.
4. Excess of glucose is converted into fats and proteins in animals.
5. Lactose synthesized in the mammary glands is a natural sweetening agent of milk of mammals.
6. Starch and glycogen are the chief food storage products in plants and animals respectively. Cellulose is the main component of the cell wall of plants.

LIPIDS

Generally, lipid means fats, but all lipids are not fats. It is an important class of biomolecules and includes a variety of compounds like oils, waxes, phospholipids and sterols in addition to fats. Chemically lipids are more diverse and are rich in elements like carbon and hydrogen but are pure in oxygen. In addition some lipids also contain traces of some other elements like nitrogen, phosphorus and sulphur. Lipids are insoluble in water but are soluble in non polar solvents like alcohol. Lipids are classified into three categories : simple lipids, conjugated lipids and derived lipids.

Simple Lipids

These are esters of alcohol and fatty acids and include fats, oils and waxes.

1. **Fats.** Fats are esters of glycerol and fatty acids and these are common in animals. Common fatty acids present in fats of animals are palmitic acid and stearic acid containing 16C and 18C respectively.

$CH_3(CH_2)_{14}COOH$ (*a*) $CH_3(CH_2)_{16}COOH$ (*b*)

Fig. 2.8. Saturated fatty acids (*a*) Palmitic acid (*b*) Stearic acid.

Glycerol is a trihydric alcohol.

$$\begin{array}{l} CH_2OH \\ \mid \\ CHOH \\ \mid \\ CH_2OH \end{array}$$

Fig. 2.9. Glycerol.

A fat is formed when saturated fatty acids react with a glycerol. Depending on the number of —OH groups of glycerol esterified a fat may be mono, di or triglyceride.

$$\begin{array}{l} H_2C\text{—}O\text{—}CO\text{—}C_{15}H_{31} \\ \mid \\ HC\text{—}O\text{—}CO\text{—}C_{15}H_{31} \\ \mid \\ H_2C\text{—}O\text{—}CO\text{—}C_{15}H_{31} \end{array}$$

Fig. 2.10. Tripalmitin (a triglyceride fat).

Tripalmitin as shown in Fig. 2.10, is a triglyceride in which all the three —OH groups of glycerol are esterified with three molecules of palmitic acid. Fats are solid at room temperature.

2. **Oils.** Oils are common in plants like mustard, sunflower, castor and groundnut. Seeds of these plants store rich amount of oils. Oils are similar to fats, the only difference is that they contain unsaturated fatty acids. When a fatty acid contains one or more double bonds then it is called unsaturated fatty acid. Three common unsaturated fatty acid containing 18C found in plants are oleic acid, linoleic acid and linolenic containing one, two and three double bonds respectively.

$$CH_3(CH_2)_7CH{=}CH(CH_2)_7COOH$$

Fig. 2.11. Oleic acid (unsaturated fatty acid).

Oils are liquid at room temperature. Triolein as shown in Fig. 2.12 is an example of oil found in plants.

$$\begin{array}{l} H_2C\text{—}O\text{—}CO\text{—}C_{17}H_{33} \\ \mid \\ HC\text{—}O\text{—}CO\text{—}C_{17}H_{33} \\ \mid \\ H_2C\text{—}O\text{—}CO\text{—}C_{17}H_{33} \end{array}$$

Fig. 2.12. Triolein oil.

3. **Waxes.** These are esters of a long chain fatty acid and a long chain monohydric alcohol. Waxes are present both in animals and plants. They are present as a surface coating on stem, leaves and fruit to check transpiration. In animals there is a coating of wax on skin and hair and is protective in nature. Due to water repelling properties of wax the feather of water birds remains dry.

Conjugated Lipids

When lipids are linked with other compounds like phosphate or oligosaccharides they are called **conjugated lipids.** Phospholipids and glycolipids are two important examples of such lipids.

1. **Phospholipids.** It is composed of three components—glycerol, fatty acids and phosphate.

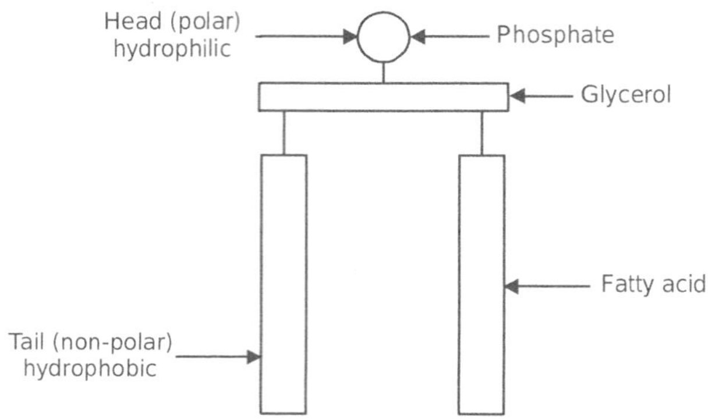

Fig. 2.13. Phospholipid (conjugated lipid).

A phospholipid molecule has two end—head and tail. Head is polar composed of phosphate and hydrophilic in nature. Tail is non-polar and is composed of long chains of hydrocarbons and it is hydrophobic in nature.

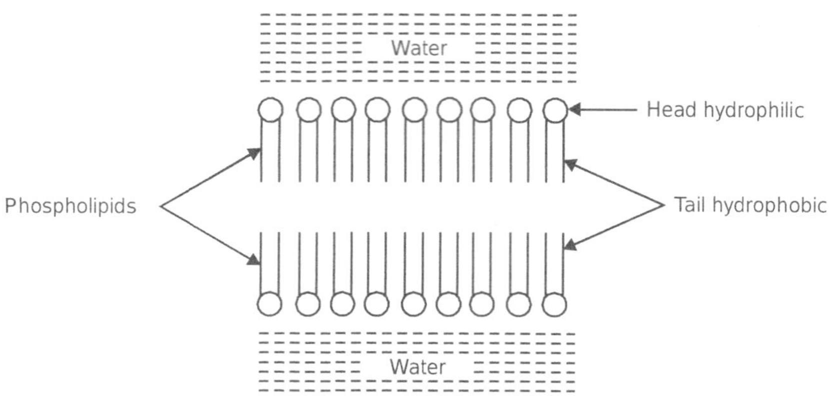

Fig. 2.14. Lipid bilayer.

Phospholipids get arranged as lipid bilayer in aqueous medium in such a way that their hydrophobic tails face the interior and hydrophilic heads on either side of the membrane as shown in the Fig. 2.14. Phospholipids constitute basic frame work of biological membrane.

2. **Glycolipids.** When lipids are linked with oligosaccharides they are called **glycolipids** and are present in the cell membrane.

Derived Lipids

It is a diverse class of lipids and contains hydrolyses product of simple and conjugated lipids, terpenes and sterols. Sterols are the compounds containing fused hydrocarbon rings and long open chain hydrocarbons. Cholesterol is an important type of sterols and is found only in animals. Cholesterol is present in blood and membrane of animals. It is a precursor for the synthesis of vitamin D and hormones like testosterone and progesterone.

Functions of Lipids

1. Lipids are one of the food storage products both in plants and animals.
2. Fats are the richest source of energy in animals and are oxidized to liberate energy when required.
3. Excess of fats in animals is deposited in adipose tissue beneath the skin and it acts as an insulator as well as shock absorber and provides protection to the internal body organs.
4. Phospholipids provide basic frame work of membrane system of cells. Glycolipids are also associated with the membranes.
5. Seeds of oil plants like, mustard, groundnut, castor, sunflower and coconut store oil which is utilized during seed germination and early embryo development.
6. Cholesterol is a precursor for the synthesis of vitamin D and hormones like testosterone and progesterone.
7. Myelin sheath around the nerve fibers acts as an electrical insulator.

PROTEINS

Protein is one of the most important bimolecules of the cell. The name protein is derived from a Greek word *proteios* means *something of primary importance.* Protoplasm of animal cell has 10 – 20% proteins. Enzymes which catalyze the biochemical reactions, hormones which regulate the metabolic process and antibodies which protect us from pathogens are all proteins. Chemically protein contain elements like C, H, O, N and some S. Some proteins also contain phosphorus and iodine and some traces of metals like iron, copper, zinc and manganese. Proteins are the macromolecules of high molecular weight and on acid hydrolysis yield amino acids.

Amino acids are the building block of proteins. A general formula of amino acid is shown in Fig 2.15. In an amino acid, four valences of carbon are fulfilled with an amino group ($-NH_2$),

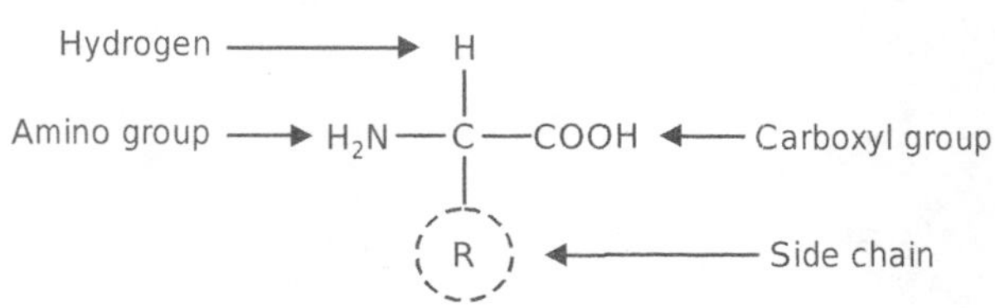

Fig. 2.15. Amino acid (general formula).

carboxyl group (COOH), hydrogen atom (H) and a side chain (R). Amino group is basic while carboxyl group is acidic in nature. There are 20 naturally occurring amino acids as shown in table 2.1 Microorganisms and plants can synthesize all 20 types of amino acids from simple inorganic nutrients. Animals including human beings depend on other sources for some of the amino acids. These amino acids are called **essential amino acids** and we obtain them from our diet.

Table 2.1 Natural amino acids

S.No.	*Name of the amino acids*	*Type of amino acids*
1.	Glycine	Neutral (aliphatic)
2.	Alanine	Neutral (aliphatic)
3.	Valine*	Neutral (aliphatic)
4.	Leucine*	Neutral (aliphatic)
5.	Isoleucine*	Neutral (aliphatic)
6.	Arginine*	Basic
7.	Lycine*	Basic
8.	Glutamic acid	Acidic
9.	Aspartic acid	Acidic
10.	Glutamine	Neutral (aliphatic)
11.	Asparagine	Neutral (aliphatic)
12.	Threonine *	Neutral (aliphatic)
13.	Serine	Neutral (aliphatic)
14.	Cysteine	Neutral (S – containing, aliphatic)
15.	Methionine	Neutral (S – containing, aliphatic)
16.	Phenylalanine*	Neutral (aromatic)
17.	Tyrosine	Neutral (aromatic)
18.	Tryptophan*	Neutral (aromatic)
19.	Histidine	Basic
20.	Proline	Neutral (aliphatic)

* Essential amino acid

Amino acids are classified as acidic, basic and neutral.

1. **Acidic Amino Acids.** In an acidic amino acid, there are two carboxyl group and one amino group for example, glutamic acid and aspartic acid.

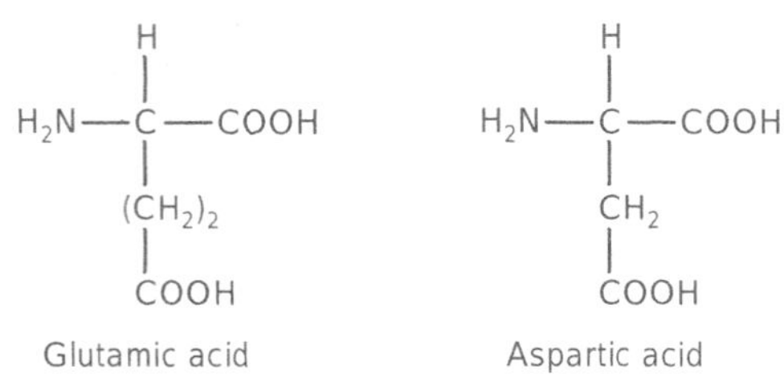

Fig. 2.16. Acidic amino acids.

2. Basic Amino Acids. In these amino acids, there is one carboxyl group and two amino groups. Arginine and lysine are two examples of basic amino acids.

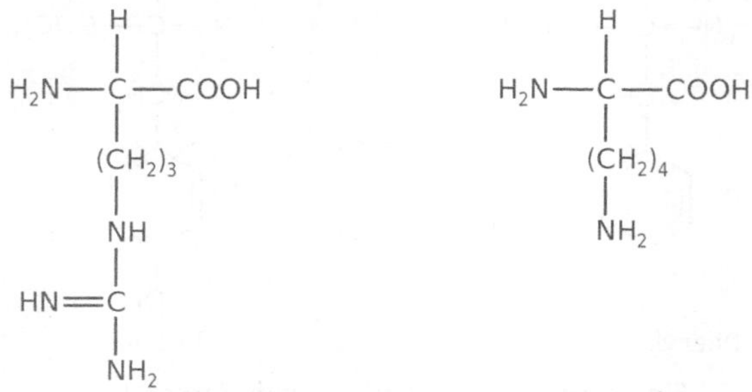

Arginine Lysine

Fig. 2.17. Basic amino acids.

3. Neutral Amino Acids. In a neutral amino acid, there is one carboxyl group and one amino group. A neutral amino acid may be aliphatic or aromatic.

(*i*) ***Aliphatic Neutral Amino Acids.*** In such amino acids, the side chain is composed of open chain hydrocarbons for example, alanine and valine.

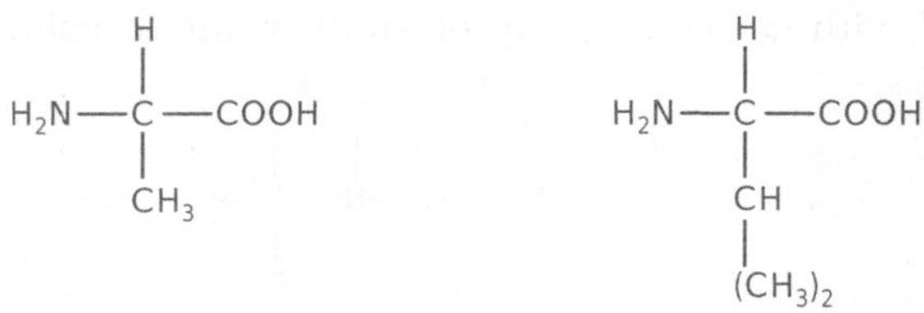

Alanine Valine

Fig. 2.18. Neutral amino acids (aliphatic).

Two of the aliphatic acids cysteine and methionine are sulphur containing.

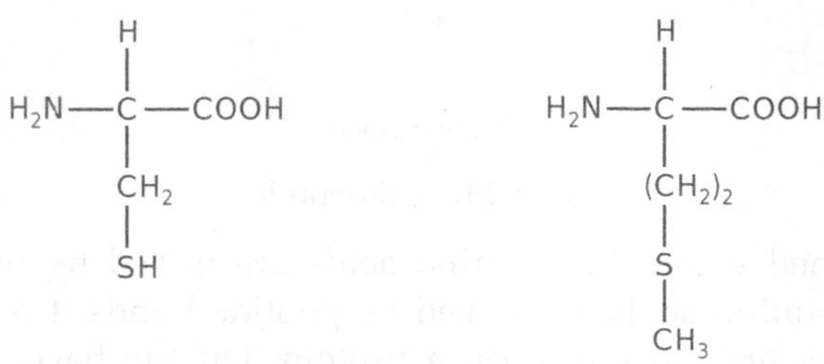

Cysteine Methionine

Fig. 2.19. Sulphur containing amino acids.

(*ii*) ***Aromatic Amino Acids.*** In the side chain of such amino acids there is at least one benzene ring for example, phenylalanine and tyrosine.

All naturally occurring amino acids except glycine are optically active.

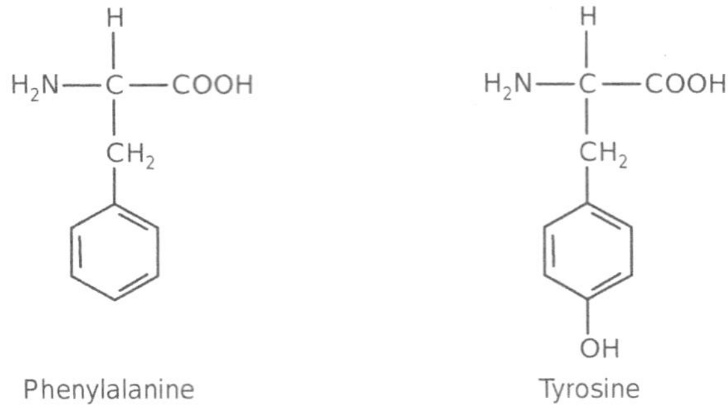

Fig. 2.20. Neutral amino acids (aromatic).

Structure of Proteins

Twenty naturally occurring amino acids get linked in various permutations and combinations to form thousands of different type of proteins. Any two amino acids can be linked by peptide bond. A peptide bond is a covalent bond which links amino group of one amino acid with carboxyl group of another amino acid. A dipeptide is formed when two amino acids get linked by a peptide bond as shown in Fig. 2.21. A peptide bond is a covalent bond which links amino group of one amino acid with carboxyl group of another amino acid.

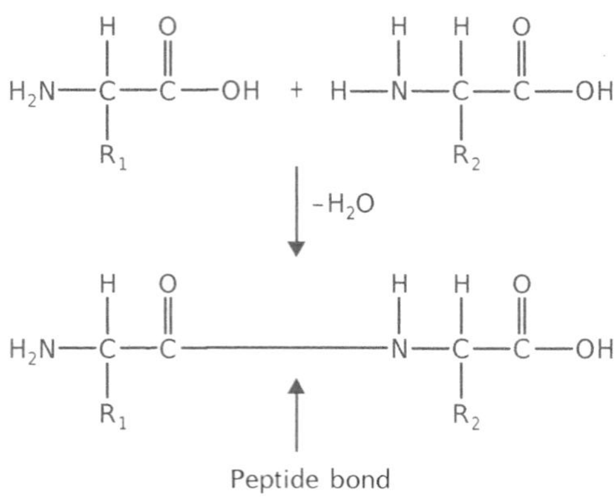

Fig. 2.21. A dipeptide.

A tripeptide is formed when three amino acids are linked by peptide bonds. Similarly, when a large number of amino acids are linked by peptide bonds it results into a polypeptide one or more polypeptides may link to form a protein. On the basis of size and complexity, protein may be primary, secondary or tertiary protein.

1. **Primary Protein.** A long polypeptide chain containing about hundred amino acids constitute a primary protein. There is a definite sequence of amino acids in a protein. Any deviation in sequencing of amino acids may altogether change the property of protein.
2. **Secondary Protein.** Secondary protein means the shape which a long polypeptide chain acquires and it is stabilized by hydrogen bonding. α helix and β pleated sheet are two commonly occurring shapes of secondary protein as shown in Fig. 2.22.

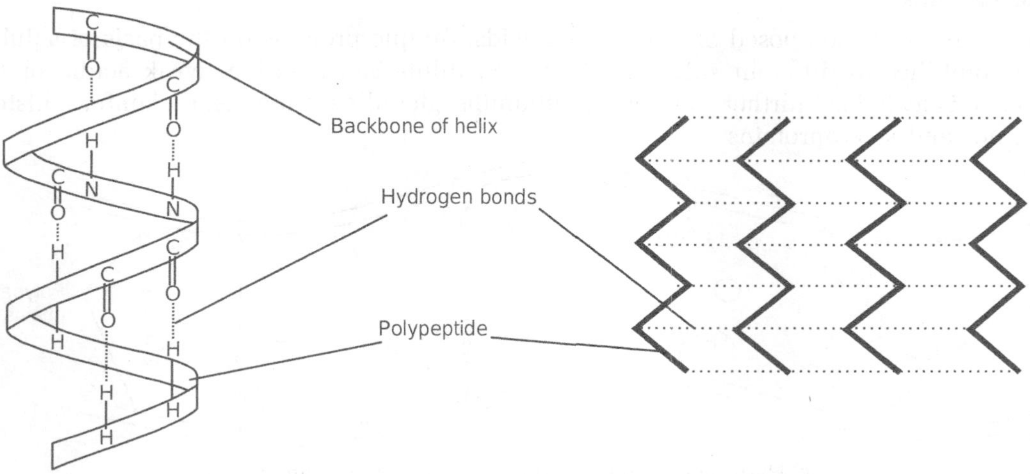

(a) α helix (b) β Pleated Sheet

Fig. 2.22. Structure of secondary proteins.

A helix conformation was first of all given by **Pauling** in 1951. In helix there is a right handed arrangement protein. There are number of hydrogen bonds between —NH_2 group of one amino acid with —C$=$O group of another amino acid of an adjacent turn. Keratin protein of hair is an important example of α helix type of secondary protein. In β pleated sheets, there are number of polypeptide chains which are arranged either by parallel or anti parallel fashion. Adjacent polypeptide chains are linked by hydrogen bonds for stabilization. Fibroin of silk fibers is an important example of β pleated sheet.

3. **Tertiary Protein.** Tertiary structure of protein is a three dimensional structure. A three dimensional structure has a definite conformation. A long polypeptide chain fold over itself or more than one polypeptide chains get arranged in some specific forms. The three dimensional structure is stabilized with the help of hydrogen bond, disulphide bond, ionic interactions and hydrophobic interaction (van der Waal forces) as shown in Fig 2.23. Tertiary protein is oriented in such a way that its non polar hydrophobic sides lie in the interior and polar and hydrophilic sides are present on the surface. Tertiary protein are functional protein *e.g.,* hormones and enzymes.

Denaturation of Protein

Tertiary protein has a definite conformation which is known as native state. Functions of such proteins are due to this specific conformation. All enzymes have a characteristic configuration and thus catalyze the biochemical reactions. Any chemical or physical change disrupts the conformation of a tertiary protein and makes it a non functional protein *e.g.,* when enzymes are subjective to high temperature (60°C) they become non functional. This is called denaturation of protein and it is an irreversible process.

Classification of Proteins

There are two main categories of proteins : Simple proteins and Conjugated proteins.

Simple Proteins

A simple protein is composed of only amino acids. Simple proteins on the basis of solubility or non solubility in different solvents like water, dilute salt solution, weak acidic or basic solution and alcohol are further classified as albumins, globulins, glutelins, prolamines, histones, protamines and scleroproteins.

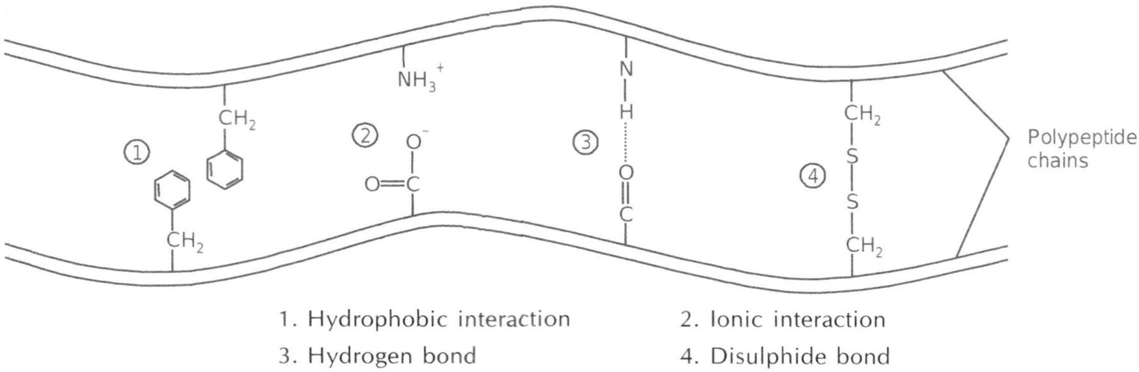

Fig. 2.23. Forces of stabilization in a tertiary protein.

(*i*) ***Albumins.*** Albumins are soluble both in water and weak salt solution. Most of the cellular enzymes like amylase, many hormones like thyroxin, egg albumin, lactoalbumin of milk and serum proteins are examples of albumins.

(*ii*) ***Globulins.*** Globulins are insoluble or sparingly soluble in water. These proteins are soluble in dilute salt solution. Some storage proteins in seeds, lactoglobulin of milk and fibrinogen found in blood plasma all are examples of globulin.

(*iii*) ***Glutelins.*** Glutelins are insoluble in water and dilute salt solution. They are soluble in weak acidic or weak basic solution. Glutelins are found in grains of cereals *e.g.*, gluenin of wheat and oryzenin of rice.

(*iv*) ***Prolamines.*** Prolamines are insoluble in water but are soluble in 70–80% alcohol. Prolamines contain large amount of proline and ammonia. They are named so because high contents of proline. Prolamines are found only in plants. Zein found in maize and gliadin found in wheat and ryes are examples of prolamines.

(*v*) ***Histones.*** Histones are water soluble proteins and are found in the nucleus of cell. Histones are rich in basic amino acids arginine and lysine. Histones are associated with DNA in nucleoproteins (chromosomes).

(*vi*) ***Protamines.*** Protamines are water soluble and are found in cell nucleus in association with DNA. Protamines are rich in basic amino acid arginine and have been reported from nuclei of spermatozoa of some fishes. Salamine and sturine are examples of protamines found in salmon and sturgeons respectively.

(*vii*) ***Scleroproteins.*** Scleroproteins are long fibrous proteins and are insoluble in water and dilute acidic or basic solution. These are found only in animals. They are found in connective tissue as well as in nails, horns, hoops and hair of animals. Important examples of scleroproteins are keratin of hair, fibroin of silk and collagen of connective tissue.

Conjugated Proteins

A conjugated protein is composed of two components, an amino component and a non amino component. Non amino component is referred as prosthetic group. Conjugated protein are named on the basis of prosthetic group *e.g.,* nucleoprotein, lipoprotein, glycoprotein, chromoprotein, mettaloprotein and phosphoprotien.

(*i*) ***Nucleoprotein.*** A nucleoprotein is composed of nucleic acid (DNA) and histone (protein). Chromatin present in the nucleus of cell is an example of nucleoprotein.

(*ii*) ***Lipoprotein.*** Lipoproteins are composed of lipids and proteins. Lipoproteins constitute the membrane system of living cells.

(*iii*) ***Glycoprotein.*** A glycoprotein is composed of carbohydrates and proteins. A glycoprotein is associated with biological membranes and acts as receptors.

(*iv*) ***Chromoprotein.*** Chromoprotein means a coloured protein. A chromoprotoein is composed of a pigment and a protein. Some examples of chromoprotiens are haemoglobin, flavoproteins, cytochromes and biliprotiens.

(*v*) ***Mettaloprotein.*** A mettaloprotein is composed of a metal and a protein. Ferritin is an iron containing mettaloprotein. Many enzymes are also associated with a prosthetic group of metal ions *e.g.,* iron (Fe^{2+}) prosthetic group of the enzyme catalase. Similarly, carbonic anhydrase enzyme contains zinc ions.

(*vi*) ***Phosphoprotein.*** A phospoprotein is composed of a phosphate and a protein. Casein present in milk is an example of phosphoprotein.

Functions of Protein

1. Proteins are an important constituent of protoplasm and are essential for growth, repair, maintenance and reproduction.
2. Proteins are structural components of biological membrane and cell organelles.
3. Some of the fibrous proteins like collagen, ligament and tendons are a part of connective tissue and act as support system which help in joining of muscles and bones.
4. Some of the proteins like actin and myosin help in muscle contraction.
5. Enzymes catalyze the biochemical reactions.
6. Hormones like insulin regulate the metabolic process.
7. Some of the body structures like nails, hair and horns etc. are protective structures in animals.
8. Antibodies protect us against pathogens.
9. Haemoglobin is a carrier of oxygen to the living cells in animals.
10. Protein acts as buffer thus regulate pH of living cells.

NUCLEIC ACIDS

Nucleic acids are the most important class of biomolecules. They are named so because of their acidic nature and were discovered from the nuclei of pus cells by **Mischer** in 1869. By the year 1940, it become clear that there are two type of nucleic acids—deoxyribose nucleic acid (DNA) and ribose nucleic acid (RNA). DNA is the genetic material of all living organisms except some viruses. On the basis of critical analysis of transformation experiment conducted by **Griffith**

(1928), in a bacterium *Diplococcus pneumoniae*; **Avery, Mac leod** and **Mc Carty** (1944), proved that DNA is the genetic material. Genetic nature of DNA was established before the discovery of its structure. DNA is present in nucleus as a constituent of chromosomes. Some DNA also present in mitochondria and chloroplast. RNA is primarily associated with the process of biosynthesis of proteins. It is also the genetic matter of some viruses like Tobacco Mosaic Virus (TMV). **Hershey** and **Chase** in 1952, proved that RNA is the genetic material in TMV. The virus that causes AIDS in human beings is also a RNA containing virus. Chemically nucleic acids are composed of C, H, O, N and P. They are consisting of three types of components which are pentose sugar, phosphoric acid and nitrogenous bases.

1. **Pentose Sugar.** Pentose sugars present in nucleotides of nucleic acids are of two types, ribose sugar ($C_5H_{10}O_5$) and deoxyribose sugar ($C_5H_{10}O_4$) present in RNA and DNA respectively. These two types of sugars are monosaccharides and their structures are shown in Fig. 2.2 in the description of carbohydrates in the beginning of this chapter. Ribose and deoxyribose differ only with respect to the attached atoms or groups on second carbon as shown in Fig. 2.2. In case of ribose there is one hydrogen atom (–H) and another hydroxyl group (–OH) at C2 whereas deoxyribose sugar has two –H attached at second carbon.
2. **Phosphoric Acid.** It is a triportic acid. It has a centrally placed phosphorous atom which is attached with three –OH group and one oxygen atom (–O) as shown in Fig. 2.24.

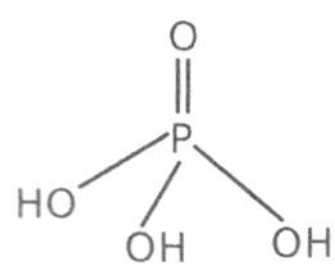

Fig. 2.24. Phosphoric acid.

3. **Nitrogen Bases.** Nitrogenous bases are heterocyclic organic compound consisting of C, H, O and N. These are of two type purines and pyrimidines as shown in Fig. 2.25.

Guanine Adenine

Purines

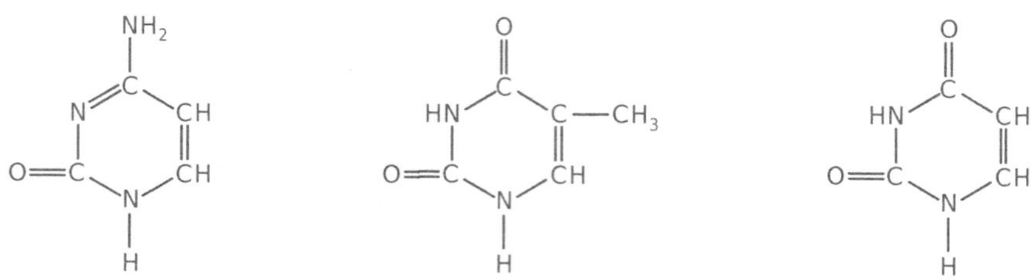

Cytosine Thymine Uracil

Pyrimidines

Fig. 2.25. Structure of nitrogenous bases.

(*i*) **Purines.** Purines are composed of two fused rings a six member ring and a five member ring. Adenine (A) and Guanine (G) are two purines.

(*ii*) **Pyrimidines.** Pyrimidines have a single six member ring. Thymine (T), Cytosine (C) and Uracil (U) are three pyrimidines. Thymine is present only in DNA while Uracil is present only in RNA and Cytosine is present in both DNA and RNA.

NUCLEOSIDE AND NUCLEOTIDE

When any of the nitrogenous bases is attached with a pentose sugar, deoxyribose or ribose the resulting structure is called a **Nucleoside**. Deoxycytidine is an example of a nucleoside which is consists of cytosine and deoxyribose sugar. A **Nucleotide** is a phosphoric ester of a nucleoside *i.e.*, it is composed of a pentose sugar, a nitrogenous base and a phosphate. Deoxycytidine monophosphate (dCMP) is a nucleotide consisting of cytosine, deoxyribose and phosphate as shown in Fig. 2.26. Three components of a nucleotide are arranged in a specific manner. A sugar is attached with the nitrogenous base at C_1 and a phosphate group is attached at C_5 as shown in the Fig. 2.26.

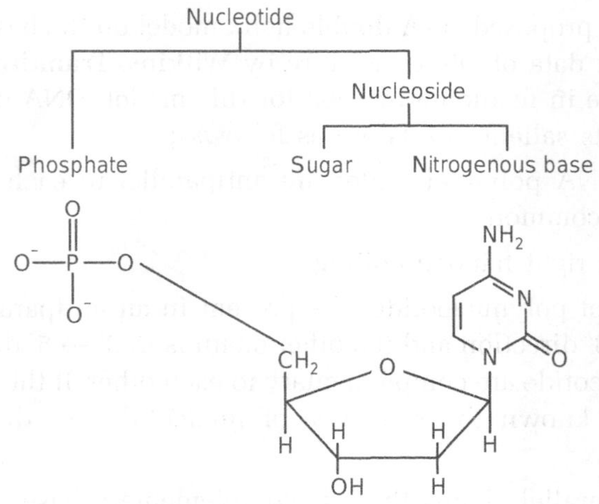

Fig. 2.26. A nucleotide (dCMP).

Types of Nucleic Acids

There are two types of nucleic acids, deoxyribose nucleic acid (DNA) and ribose nucleic acid (RNA).

1. Structure of DNA

A DNA is composed of two very long chains of polynucleotides. A nucleotide of DNA is consists of a deoxyribose sugar, phosphate and any of the following four nitrogenous bases : Adenine (A), Guanine (G), Thymine (T) and Cytosine (C). DNA has four types of nucleotides which are dAMP, dGMP, dCMP and dTMP as shown in table 2.2.

Table 2.2 Nucleotides of DNA

S.No.	*Name of the nucleotide*	*Abbreviation*	*Composition*
1.	Deoxyadenosine monophosphate	dAMP	Adenine + deoxyribose + phosphate
2.	Deoxyguanosine monophosphate	dGMP	Guanine + deoxyribose + phosphate
3.	Deoxythymidine monophosphate	dTMP	Thymine + deoxyribose + phosphate
4.	Deoxycytidine monophosphate	dCMP	Cytosine + deoxyribose + phosphate

Fundamentally the structure of all the nucleotides of DNA is same and a detailed structure of a nucleotide dCMP (deoxycytidine monophosphate) is given in Fig. 2.26. The only difference is in the type of nitrogenous base as shown in table 2.2. **Chargaff** (1950), analyzed DNA of different living organisms and concluded that proportion of purines is always equal to pyrimidines. He also concluded that

$$A + T \neq G + C$$

DNA Double Helix Model

Watson and **Crick** (1953), proposed DNA double helix model on the basis of Chargaff's findings and x-ray crystallography data of DNA structure by **Wilkins, Franklin** and **co workers.** They were awarded Noble prize in medicine in 1962 for this model. DNA double helix structure is shown in Fig. 2.27 and its salient points are as follows :

1. Two chains of DNA polynucleotides are antiparallel to each other and are helically coiled around a common axis.
2. DNA coiling is a right handed coiling.
3. The two chains of polynucleotides are present in an antiparallel manner. One of the chain is in $5' \rightarrow 3'$ direction and the other chain is in $3' \rightarrow 5'$ direction. The two chains of DNA polynucleotide are complementary to each other. If the sequence of nucleotides of one chain is known the sequence of nucleotides in the second chain can be predicted.
4. In the two antiparallel chains there is complementary base pairing. Adenine always pair with thymine by two hydrogen bonds similarly guanine always pair with cytosine by three hydrogen bonds. The two chains are stabilized by hydrogen bonds between complementary bases. A polynucleotide chain is formed by linking different DNA nucleotides in a random manner.

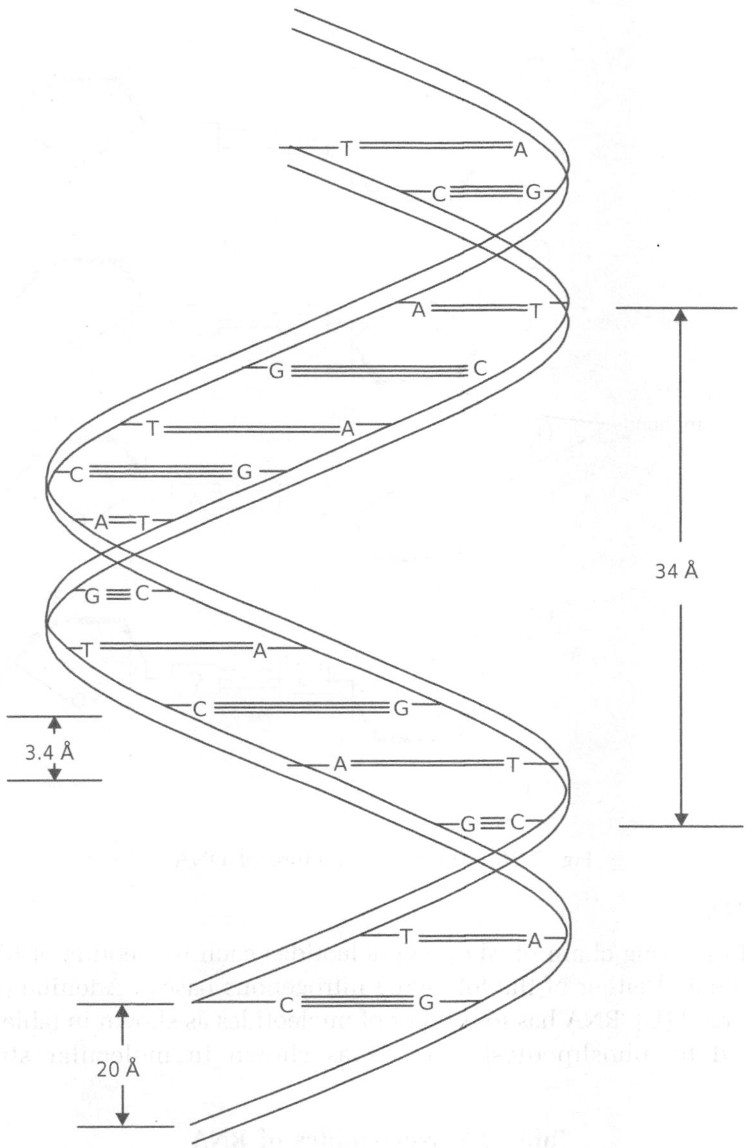

Fig. 2.27. Watson Crick model of DNA double helix.

5. The linkage between adjacent nucleotides is of ester type. The adjacent nucleotides are linked by phosphodiester bonds. A phosphodiester bond is formed between 5′ and 3′ hydroxyl groups of two deoxyribose sugars of successive nucleotides as shown in the molecular structure of DNA (Fig. 2.28).
6. The two strands are at a constant distance of 20 Å.
7. The distance between two successive turns of coil is 34 Å and contains ten base pairs per turn.
8. The distance between two successive nucleotides of a polynucleotide chain is 3.4 Å.

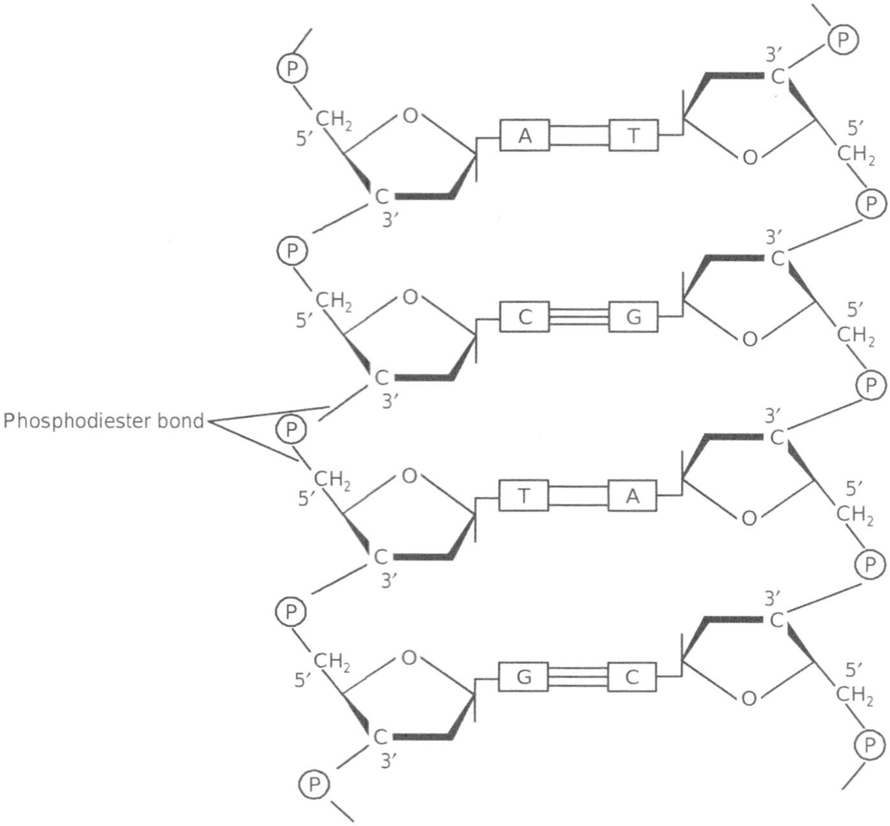

Fig. 2.28. Molecular structure of DNA.

2. Structure of RNA

RNA is composed of a long chain of ribopolynucleotides each nucleotide of RNA is composed of ribose, phosphate and either of the following nitrogenous bases : Adenine (A), Guanine (G), Cytosine (C) and Uracil (U). RNA has four types of nucleotides as shown in table 2.3. Nucleotides of RNA are linked by phoshpodiester bonds as shown in molecular structure of RNA (Fig. 2.29).

Table 2.3. Nucleotides of RNA

S.No.	*Name of the Nucleotide*	*Abbreviation*	*Composition*
1.	Adenosine monophosphate	AMP	Adenine + ribose + Phosphate
2.	Guanosine monophosphate	GMP	Guanine + ribose + Phosphate
3.	Cytidine monophosphate	CMP	Cytosine + ribose + Phosphate
4.	Uridine monophosphate	UMP	Uracil + ribose + Phosphate

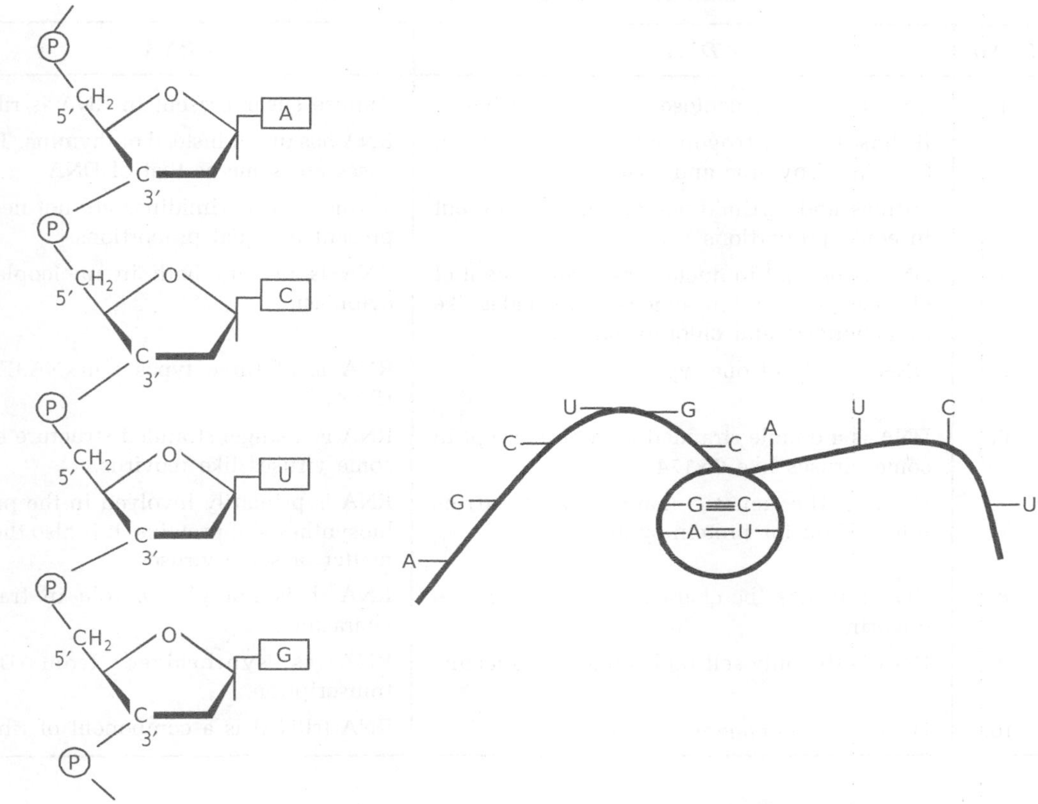

(*a*) **Molecular structure of RNA** (*b*) **A single strand of RNA**

Fig. 2.29. Stucture of RNA.

Unlike DNA, RNA is a single stranded structure (Fig. 2.29) and it differs from DNA in many respects as shown in table 2.4. RNA does not have complementary base pairing as in DNA. However, the long strand of RNA occasionally forms some loops which have a few paired bases. Adenine pair with uracil and cytosine pair with guanine by two and three hydrogen bonds respectively. RNA is synthesized with the help of DNA by a process known as **transcription.** RNA is of three types : messenger RNA, transfer RNA and ribosomal RNA.

(*a*) **Messenger RNA (mRNA).** It is a small sized RNA consisting of about 900–1500 nucleotides and constitutes 5–10% of total cellular RNA. It carries information from DNA to the site of protein synthesis in the cytoplasm and the term RNA was coined by **Jacob** and **Monod** (1961). mRNA represents a pattern that controls the synthesis of proteins.

(*b*) **Transfer RNA (tRNA).** It is the smallest RNA containing about 70–80 nucleotides. It comprises 10–15% of cellular RNA. It carries amino acids from the cellular pool of amino acids present in cytoplasm to the site of protein synthesis.

(*c*) **Ribosomal RNA (rRNA).** It is of a very large size and contains thousands of nucleotides. It comprises about 80% of total cellular RNA. rRNA is a component of ribosome.

Table 2.4. Difference between DNA and RNA

S. No.	*DNA*	*RNA*
1.	DNA contains a pentose sugar deoxyribose.	Pentose sugar present in RNA is ribose.
2.	It has four nitrogenous bases Adenine, Guanine, Thymine and Cytosine.	RNA has uracil instead of thymine. The other bases are same as that of DNA.
3.	Purines and pyrimidines are always present in equal proportions.	Purines and pyrimidines are not necessarily present in equal proportions.
4.	DNA is present in nucleus as a component of chromosomes and in some cell organelles like mitochondria and chloroplast.	RNA is present both in nucleoplasm and cytoplasm.
5.	DNA is only of one type.	RNA is of three types : mRNA,tRNA and rRNA.
6.	DNA is a double stranded structure except in some viruses like Φx174.	RNA is a single stranded structure except in some viruses like reovirus.
7.	DNA is the genetic material and carries information for protein synthesis.	RNA is primarily involved in the process of biosynthesis of proteins. It is also the genetic matter of some viruses.
8.	DNA transfers the characters from parents to offspring.	RNA does not play a role in transfer of characters.
9.	DNA is the only self replicating biomolecule.	RNA is synthesized from DNA by transcription.
10.	DNA is a component of chromosomes.	RNA (rRNA) is a component of ribosomes.

Functions of Nucleic Acids

DNA is the genetic material of living organisms which governs the characters of living organisms. It is also the hereditary matter which transfers characters from parents to the offspring.

1. Nucleic acids both DNA and RNA play very significant role of biosynthesis of proteins in living organisms.
2. DNA play a role in variations and as well as in the organic evolution.

VITAMINS

Vitamins are the organic compounds which are required in minute amounts for normal growth, development and maintenance of vigour by all the living organisms. Most of the vitamins act as components of coenzymes and help in catalyzing the biochemical reactions. **Funk** (1912), detected the presence of certain substances in the food of animals which are essential for their normal metabolic activities and named these substances as vitamins. Vitamins are named using English alphabets A, B, C, D, E, etc., on the basis of chronology of their discovery. Some vitamins also have sub groups. The names of sub groups are written by putting suffix 1, 2, 3, etc., after the name of parent vitamin *e.g.,* Vitamins B_1, B_2, B_3, etc., are the names of sub groups of vitamin B_{12} complex. These trivial names of vitamins are still in common usage though the chemical names of most of the vitamins are available *e.g.*, retinol, thiamine, ascorbic acid are the names of vitamins A, B_1 and C respectively.

Plants can synthesize all the vitamins. Animal including human beings can synthesize only some of the vitamins. Requirements of other vitamins are met with the intake of dietary component by animals. Our body requires vitamins A, B, C, D, E, and K. Some of these vitamins are synthesized in our body or made available to us by the bacteria present in our alimentary canal. Vitamin D is synthesized in the presence of sunlight. The ultraviolet rays of sunlight converts sterols present in body tissues into vitamin D. Some of the vitamins of B_{12} complex and vitamin K are synthesized by the microbes present in the alimentary canal and are absorbed for utilization. If vitamins are not made available to us in required amounts they cause deficiency diseases as shown in table 2.5

Table 2.5. Deficiency diseases of vitamins

Name of Vitamin	*Deficiency Diseases*
Vitamin A (retinol)	Night blindness and xerophthalmia
Vitamin B_1 (thiamine)	Beri-beri
Vitamin B_2 (riboflavin)	Dermatitis and lip sores
Vitamin B_6 (pyridoxine)	Severe dermatitis and convulsions
Vitamin B_{12} (cyanocobalmin)	Pernicious anemia
Vitamin C (ascorbic acid)	Scurvy
Vitamin D (calciferol)	Rickets
Vitamin E (tocopherol)	Sterility and muscular dystrophy
Vitamin K (1, 4-napinthoquinone acetate)	Haemorrhage

On the basis of solubility, vitamins are classified either as water soluble or lipid soluble. Vitamin B and C are water soluble and vitamins A, D, E and K are lipid soluble. A brief account of role of different vitamins in the physiological process, their natural sources of availability and the effect of their deficiency in humans is given below :

1. **Vitamin A.** It is also known as retinol. It is important for healthy eyes. It is required for proper growth and maintenance of mucous membrane and glands of eyes. It also acts as precursor in the synthesis of light absorbing pigment rhodopsin of rods cells of eye retina. Natural sources of vitamin A are leafy vegetables, carrots, egg yolk, butter and fish liver oil. Deficiency of vitamin A causes night blindness *i.e.*, defective vision in dim light and xerophthalmia in which mucus membrane of eye is weakened and cornea becomes dry.
2. **Vitamin B.** It is a group of 12 vitamins and plays varied role in living organisms. Some of these vitamins like B_1 (thiamine), B_2 (riboflavin) and B_3 (niacin) are components of coenzymes NAD, NADP and FMN. These coenzymes carry out the oxidation-reduction reactions of cellular respiration. Similarly, vitamin B_6 (pyridoxine) is a part of coenzyme required for the synthesis of amino acids and glycogen. Vitamin B_{12} is needed for the formation of blood cells.

Vitamin B_{12} are water soluble and are present in grain of cereals, pulses, carrot, cabbage, green vegetables, nuts, milk, meat and yeast. Deficiency of these vitamins causes a number of diseases like beri beri, skin disorder and anemia. Deficiency of vitamin B_1 causes beri beri. It is due to consumption of polished rice. In beri beri body muscles are weakened and in severe cases it may lead to paralysis. Deficiency of vitamin B_2 impairs growth and causes dermatitis which is a skin disorder. Deficiency of vitamin B_{12} causes pernicious anemia.

3. **Vitamin C.** It is also known as ascorbic acid. It is obtained only from external resources human body cannot synthesize it. Vitamin C is a reducing agent and participates in many oxidation-reduction reactions of the body. It is presumed to resist infection and stress. It is very essential for normal metabolism and is also required for normal growth of bones and teeth. Citrus fruits, tomatoes and leafy vegetables are good sources of vitamin C, it is a thermo labile vitamin which is destroyed on heating. Deficiency of vitamin C causes scurvy which is characterized by bleeding gums, painful and swollen joints and delay in healing of wounds.
4. **Vitamin D.** It is a lipid soluble vitamin and is also known as calciferol. It helps in absorption of phosphorus and calcium by the gut. Proper amount of phosphorus and calcium is must for the normal growth of bones and teeth. Vitamin D is synthesized in human body in sunlight while other important sources of this vitamin are fish liver oil, yeast and egg yolk. Deficiency of vitamin D causes, rickets in children which results in abnormal teeth and bow bones. In adults, its deficiency causes osteomalacia in which bones become weak and are prone to fractures.
5. **Vitamin E.** It is a lipid soluble vitamin and is also known as tocopherol. It is essential for rapid cell growth and reproduction in animal. Some of the natural resources are cereals, leafy vegetables, milk and butter. Its deficiency causes sterility.
6. **Vitamin K.** It is a lipid soluble vitamin. Its chemical name is 1, 4-napinthoquinone acetate. It helps in the formation of blood protein prothrombin. It is required to stop bleeding during injury to form a clot. So it prevents flow of excess blood from the body. It is synthesized by alimentary canal bacteria. Green vegetables, soybean oil and tomatoes are good sources of vitamin K. Its deficiency causes haemorrhage.

Exercises

1. Classify carbohydrates and give a brief account of monosaccharides.
2. Describe the food storage polysaccharides ?
3. Describes the composition and function of cellulose ?
4. What are disaccharides ? Give suitable examples.
5. What are reducing sugars ?
6. Enumerate different functions of carbohydrates ?
7. Name the sweetest naturally occurring sugar.
8. What are lipids ? Give a brief account of simple lipids.
9. Differentiate between a fat and oil.

10. What are the functions of lipids ?

11. What are proteins ? Describe secondary proteins with suitable examples ?

12. Write a brief note on tertiary protein.

13. What do you mean by denaturation of proteins ?

14. What are essential amino acids ? Name any two.

15. What are nucleic acids ? What are different components of nucleic acid ?

16. What are a nucleoside and a nucleotide ?

17. Describe Watson Crick double helix DNA model with the help of well labeled diagram.

18. Differentiate between DNA and RNA.

19. Write a brief note on vitamin B_{12} complex.

20. Name the deficiency diseases due to lack of vitamin A, B_1, B_{12}, C, D, E and K.

❑❑❑

Chapter 3 ENZYMES

Enzymes are the organic compounds mainly composed of proteins and work as biocatalysts. A biocatalyst is similar to a catalyst in its action which enhances the rate of a reaction. At a given point of time, hundreds of biochemical reaction takes place in cells of living organisms. Each of these reactions is catalyzed by an enzyme. Biochemical reactions may occur in the absence of enzymes but at a very slow rate; at such a slow rate that existence of life is practically not possible. Enzymes are synthesized inside a cell by its DNA. DNA of each cell has necessary information to synthesize different enzymes as and when required. Though enzymes are secreted inside the living cells they retain their catalytic properties when extracted from the cells. One such enzyme used commercially is rennin extracted from calf's stomach and is used to convert milk into cheese. Fermentation, a type of biochemical reaction in which sugar is converted into alcohol and carbon dioxide is utilized for the production of wine and bread since time immemorial.

Enzymes were discovered accidentally by **Buchner** in 1897. He prepared a yeast extract for some medicinal purpose. He added some sugar solution in it to check its spoilage. He noticed after a few days that the container in which yeast extract was preserved on opening gave an alcoholic smell. Later, he confirmed that fermentation had occurred in it. **Pasteur,** a well known microbiologist already noticed that fermentation occurs due to living yeast cells. Buchner reported that fermentation is also possible in a cell free environment. He reported that factors present in yeast extract bring about fermentation and thus the term enzyme means 'in yeast' was coined. **Sumner** (1926), first of all crystallized an enzyme urease and reported that it is composed of protein and he also postulated that all enzymes are made up of proteins.

CHARACTERISTICS OF ENZYMES

Enzymes act as a biocatalyst, catalyzing various biochemical reactions. Enzymes resemble inorganic catalysts as both enhance the rate of reactions, still they differ with each other in many ways as shown in table 3.1. Some important characteristics of enzymes are as follows :

1. Chemical Nature

Enzymes are the proteinaceous compound and has a characteristic three dimensional structure. Catalytic property of an enzyme is by virtue of one or more active sites that it possesses. Some enzymes are only made up of proteins *e.g.*, amylase, lipase, urease etc. Many enzymes have a non protein component called a prosthetic group which is essential for its activity. In such cases protein component is known as **apoenzyme.** Apoenzyme plus prosthetic group together form **holoenzyme.** Apoenzyme in isolation cannot catalyse a reaction. A tightly bound prosthetic composed of a metal compound iron-porphyrin is associated with cytochromes. Non protein component in many enzymes is loosely associated. It may be a metal ion or an organic compound. It is called an inorganic cofactor when composed of metals or their ions. When it is composed of an organic molecule it is called organic cofactor (coenzyme). A brief description of coenzymes and cofactors is given at the end of this chapter.

2. Fast in Action

Enzymes are very fast in action. In the presence of enzymes biochemical reactions occur thousand or million times faster than without enzymes. The enzyme carbonic anhydrase present in red blood cells helps in the formation of carbonic acid in animal tissues by combining carbon dioxide with water for the removal of CO_2 from the body through lungs. Carbonic anhydrase is the fastest known enzyme. Its turnover number is 36 million; it means that one mole of enzymes carbonic anhydrase converts 36 million moles of carbon dioxide into as many moles of carbonic acid. Turnover number of enzymes varies from 10^2 to 3.6×10^6.

3. Specificity

The action of enzymes is highly specific. An enzyme catalyses only one type of reaction not others. For example, enzyme catalase can break hydrogen peroxide (H_2O_2) into water (H_2O) and oxygen (O_2).

$$H_2O_2 \xrightarrow{\text{Catalase}} H_2O + 1/2O_2$$

Catalase cannot break a maltose molecule into its constituents. Maltose ($C_{12}H_{22}O_{11}$) can be hydrolyzed into its constituents *i.e.*, two units of glucose ($C_6H_{12}O_6$) only by the enzyme maltase.

$$C_{12}H_{22}O_{11} + H_2O \xrightarrow{\text{Maltase}} 2C_6H_{12}O_6$$

4. Sensitive to Temperature and pH

Enzymes are sensitive to change in both temperature and pH. At high temperature $< 60°C$ enzymes get denatured *i.e.*, become non functional, they are also sensitive to pH. Every enzyme works in a very narrow range of pH.

5. Remains Unaltered

An enzyme after catalyzing a reaction remains unaltered *i.e.*, it is not affected in any way during its functioning after the completion of a reaction it can be reused. So an enzyme has the same shape and size before initiating or after completing the reaction.

6. Reversibility of Reaction

Although enzymes increase the rate of reaction they do not alter the equilibrium of a reversible biochemical reaction. Enzymes only help to reach equilibrium state much faster. In the absence of an enzyme a reaction will proceed towards equilibrium at a very slow speed.

Table 3.1. Difference between enzymes and inorganic catalyst

S.No.	*Enzymes*	*Inorganic Catalyst*
1.	Enzymes are organic compounds made up of proteins which catalyses the biochemical reactions in living organisms.	Catalysts are inorganic compounds consisting of metals like iron, nickel, cobalt, molybdenum, zinc, manganese etc which catalyses simple chemical reactions.
2.	Enzyme catalyses the reaction at normal body temperature.	Inorganic catalyst catalyses the reactions usually at very high temperature or high pressure or both.
3.	Enzymes are very specific in nature an enzyme catalyses only one type of reaction.	Inorganic catalyst catalyses a number of chemical reactions they are not very specific in their action.
4.	Enzymes are very sensitive to change in temperature and pH. At very high temperature enzymes become non functional. Similarly, they work in a narrow range of pH.	They are not so sensitive to change in temperature and pH.
5.	Activity of enzymes is regulated by some regulator molecules. The regulator molecules regulate the enzyme activity by changing their conformation.	Inorganic catalysts are not regulated by any regulator molecules.

MODE OF ENZYME ACTION

Enzymes carry out biochemical reactions as they have some active sites on their surface. These active sites help in making a reaction possible in a very short duration of time. There are two hypothesis to explain mode of enzyme action. These are :

1. Lock and Key Hypothesis

This theory was proposed by **Fisher** (1898), according to this theory *every enzyme has a one or more fixed active sites.*

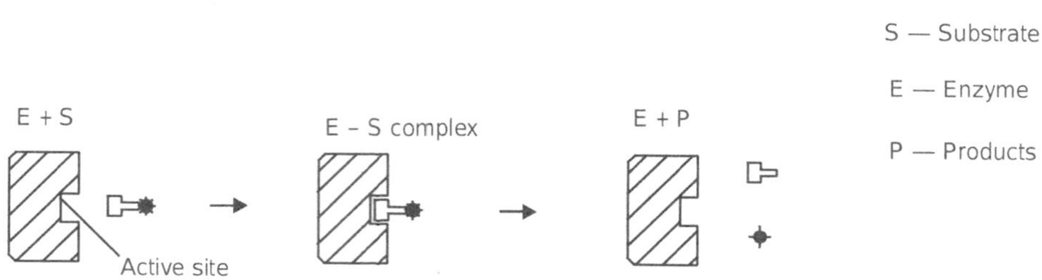

Fig. 3.1. Diagrammatic representation of Lock and key hypothesis.

An active site has a specific geometric configuration. In an active site only a specific substrate can fit in to bring out a biochemical reaction as shown in Fig 3.1 in an active site a substrate is bounded by some weak forces like hydrogen bonds. In this way an enzyme substrate complex is formed temporarily. Both the components of enzyme substrate complex are believed to have strictly complementary structures. Active sites also promote a chemical reaction and release the product after the release of the product the enzyme becomes free to carry out the same type of reaction again. This hypothesis is analogous to the fact that a particular lock only accepts a key of a specific configuration.

2. Induced Fit Hypothesis

According to this hypothesis active site of an enzyme is not a rigid structure. Some adjustments are made in the conformation of the active site when a specific substrate approaches it. This theory was proposed by **Koshland** in 1959. According to this theory enzyme and substrate have a 'hand and glove' relation as hand fits into a glove in the similar fashion a specific substrate fits in the active site of an enzyme to carry out a biochemical reaction.

MECHANISM OF A BIOCHEMICAL REACTION

To understand mechanism of a biochemical reactions consider a generalized reaction.

$$E + S \longrightarrow ES \longrightarrow E + P$$

As shown in this reaction a substrate (S) is converted into a product (P) with the help of an enzyme (E). In this reaction first an enzyme and a substrate integrate to form an enzyme substrate (ES) complex. Finally the enzyme substrate complex releases the product and enzyme becomes free. In different biochemical reactions a substrate may be one or more. Similarly there may be one or more product. An enzyme lowers the energy of activation and hence increases the rate of reaction as shown in Fig. 3.2.

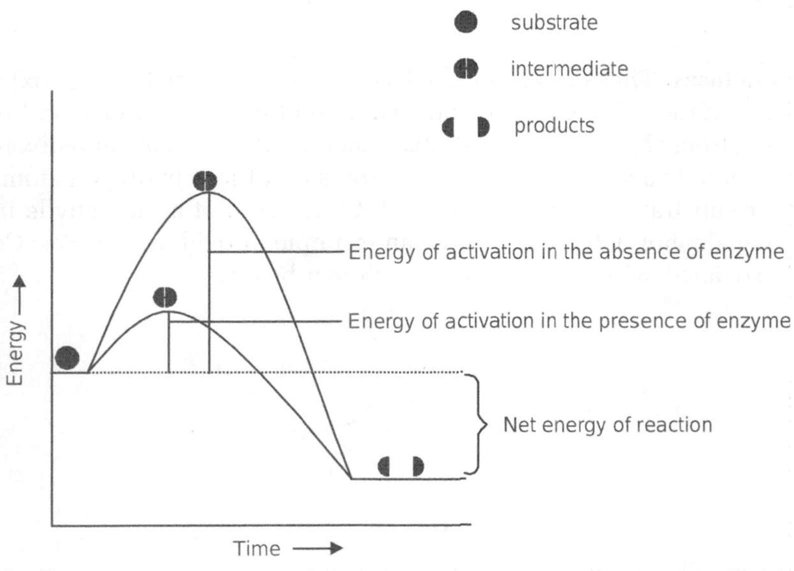

Fig. 3.2. Mechanism of enzyme action.

In any biochemical reaction a substrate needs some energy called **energy of activation** to react. During the process first a transition state is achieved. Transition state is achieved faster and it requires less energy in the presence of enzyme than in its absence. Intermediate transition state is transformed into products. The difference between energy level of substrate and product is called net energy of reaction. In the absence of enzymes a biochemical reactions needs much higher energy *e.g.*, in the breakdown of hydrogen peroxide into water and oxygen requires 18 k Cal mol^{-1} whereas in the presence of enzyme catalase required energy is just 6.4 k Cal mol^{-1} which is available at normal body temperature.

NOMENCLATURE AND CLASSIFICATION OF ENZYMES

Enzymes are generally named by adding a suffix—ase after the name of a substrate on which it acts. For example sucrase is the name of an enzyme that breaks sucrose into glucose and fructose. Similarly amylase, protease, lipase, are the names of some enzymes which hydrolysis substrates like amylose (starch), proteins and lipids respectively into their smaller components. Name of the enzymes also indicates the type of reaction they catalyses. The above mentioned enzymes are involved in those reactions in which bigger molecules are broken into smaller molecules, similarly enzyme ATP synthase helps in the synthesis of ATP from ADP and phosphate. Some of the enzymes are named arbitrarily. Trypsin and pepsin are two such enzymes having arbitrary names. Such trivial names are still retained because of their common usage. Enzymes are grouped into six classes :

1. Oxidoreductases
2. Transferases
3. Hydrolases
4. Lyases
5. Isomerases
6. Ligases

1. **Oxidoreductases.** These enzymes catalyses oxidation-reduction (redox) reactions. In a redox reaction there is a simultaneous oxidation (loss of electrons) and reduction (gain of those electrons) by another substrate. Generally in biochemical redox reaction during oxidation there is also loss of hydrogen atoms and these hydrogen atoms are accepted by another substrate which gets reduced. Conversion of acetaldehyde into ethanol by the enzyme alcohol dehydrogenase is an example of oxidoreductases. Coenzyme NAD is also associated with this reaction as shown below.

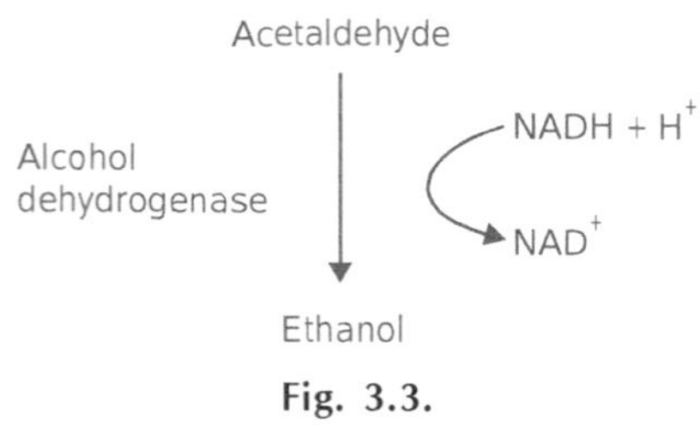

Fig. 3.3.

2. **Transferases.** These enzymes catalyzes those reaction in which some specific groups are transferred from one substrate to another substrate *e.g.*, enzyme glutamic-aspartic transaminase transfers amino group from glutamic acid to oxaloacetic acid to form aspartic acid.
3. **Hydrolases.** These enzymes break a macromolecule into small molecules by breaking

covalent bonds by the addition of water molecules. Important examples of such enzymes are digestive enzymes like amylase, protease and lipase as already mentioned in the nomenclature of enzymes.

4. **Lyases.** These are enzymes which break the bond of a big molecule to form small molecules without addition of water molecule. Oxalosuccnic acid decarboxylase is an example of lyases which breaks oxaloacetic acid into α-ketoglutaric acid and carbon dioxide.

5. **Isomerases**. These enzymes help in converting a substrate into its structural isomer *e.g.*, phosphoglucoisomerase converts glucose 6-phosphate into fructose 6-phosphate during glycolysis.

6. **Ligases.** These enzymes help in joining two or more small molecules to form a big molecule *e.g.*, ATP synthatase helps in forming an ATP molecule from ADP and inorganic phosphate. Similarly DNA ligase helps in linking small fragments (Okazaki fragments) of DNA during the process of DNA replication.

FACTORS AFFECTING ENZYME ACTIVITY

Role of enzymes in biochemical reactions is to speed up the reaction. As enzymes are sensitive to change in the internal as well as external environment, so the rate of reaction is influenced by many factors like change in substrate and enzyme concentration, temperature and pH. As shown with the help of graphs in Fig. 3.4.

1. Substrate Concentration

Enzymes activity is less at very low substrate concentration. If substrate concentration is increased gradually, activity of enzyme also increases. Finally a point is reached at which further increase in substrate concentration does not increase the enzyme activity. The rate of reaction at this stage becomes constant as shown in Fig. 3.4 (*a*). This is because of enzyme saturation effect. It means all the available molecules of enzymes are busy in catalyzing the reaction so at this point for the further added substrate availability of enzymes become a limiting factor and hence reaction rate becomes constant.

2. Enzyme Concentration

If substrate concentration is more and enzyme concentration is very low, rate of reaction is slow. This is because there are only a few molecules of enzymes which can convert limited amount of substrate into product hence enzyme activity is less. Increase in concentration of enzymes increases the rate of reaction sharply as shown in Fig. 3.4 (*b*).

Increase in concentration of enzyme results in conversion of more and more substrate into products. So rate of reaction always increase sharply at high enzyme concentration.

3. Temperature

Effect of temperature on an enzyme catalyzing reaction is shown in Fig 3.4 (*c*). The optimal range of temperature for most of the enzymes lies between 25° – 40°C. Optimal temperature means at this temperature enzyme activity is at its best. Human enzymes work at best at body temperature *i.e.*, 37°C. At very low temperature > 5°C enzyme activity is negligible. Every increase in 10° rise in temperature almost doubles the rate of reaction. Increase in temperature after 45°C decreases the rate of reaction and at 60°C it becomes zero. This is due to the reason that at such high temperature enzymes get denatured and hence become nonfunctional. Though thermophilic archaebacteria can tolerate < 80°C, it is due to their adaptability at very high temperature.

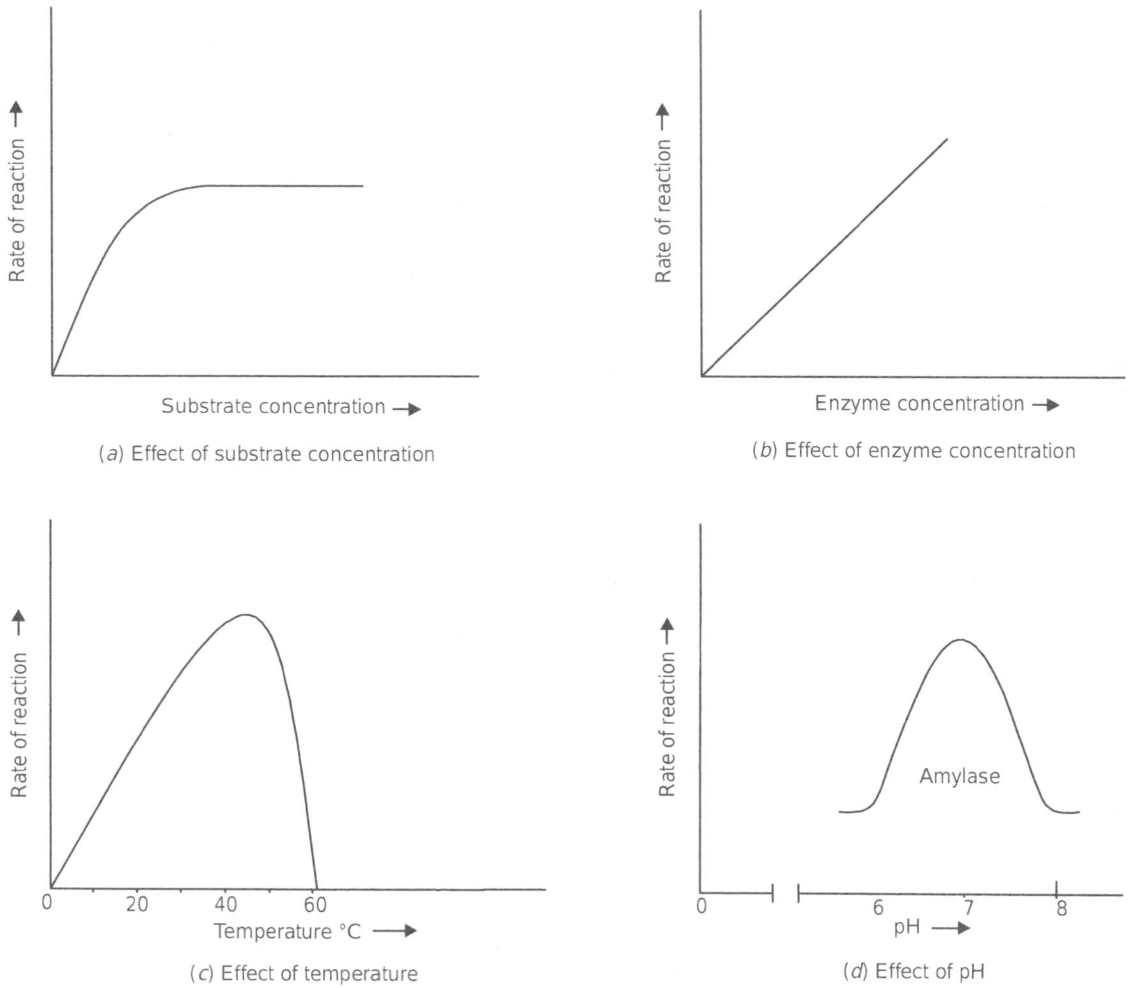

Fig. 3.4. Factors affecting enzyme activity.

4. Hydrogen Ion Concentration (pH)

Enzymes are sensitive to change in pH. Different enzymes work at different optimal pH. Some enzymes work in acidic pH, some in neutral pH and others in alkaline pH. Effect of change in pH on the activity of enzymes amylase is shown in 3.4 (*d*). As it is clear from the diagram that amylase work in a narrow range of pH and its optimal pH is 6.7. Similarly two protein digesting enzymes trypsin (in stomach) and pepsin (in duodenum) are active at pH 2 and pH 8.8 respectively. Any deviation from the optimal pH decreases the enzyme activity.

INHIBITION OF ENZYME ACTIVITY

Enzyme activity may be inhibited by some compounds known as inhibitors. Enzyme activity is inhibited by three ways : Competitive inhibition, Non competitive inhibition and Feedback inhibition

1. Competitive Inhibition

In case of competitive inhibition an inhibitor competes for the active site of an enzyme with its substrate. This occurs due to resemblances in the structure of a substrate and an inhibitor. When an inhibitor occupies the active site of an enzyme then an enzyme—inhibitor complex is formed in which no further action takes place. It is analogous to a situation of getting a lock jammed by a fake key. Competitive inhibition supports lock and key hypothesis of enzymes mode of action. For example, enzyme succinic acid dehydrogenace catalyzes the convertion of succinic acid into fumaric acid. Sometimes, malonic acid which resemble succinic acid in structure occupies the active site of the enzyme and no reaction takes place as shown in Fig. 3.5.

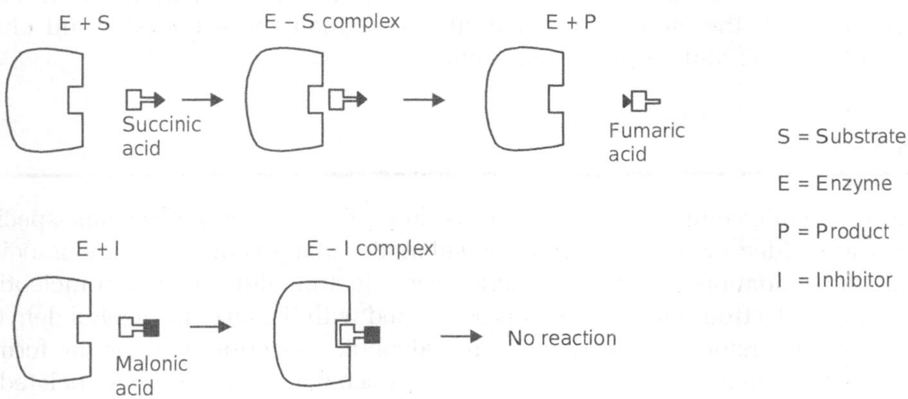

Fig. 3.5. Competitive inhibition of enzyme activity.

Competitive inhibition can be overcome by increasing the substrate concentration. By the same principle of competitive inhibition sulpha drugs controls pathogenic bacteria as they compete with the substrate *p*-amino benzoic acid and stops the synthesis of folic acid which is an important metabolite required by the bacteria.

2. Non Competitive Inhibition

In case of non competitive inhibition the inhibitor does not occupies the active site of the enzyme. Instead it attaches with the enzyme at some point and thus changes the conformation of the enzyme making it non functional. Death due to cyanide poisoning in animals is an example of this type of inhibition. Cyanide attaches with the cytochrome oxidase, a mitochondrial enzyme essential for cellular respiration at some site other than the active site and blocks the functioning which causes death.

3. Feedback Inhibition

Feedback inhibition means inhibiting the activity of an enzyme by some intermediate compound synthesized in those biochemical reactions which takes place by a number of steps and each step is catalyzed by a specific enzyme. If the final product is no more required then enzyme activity is inhibited at some intermediate step. For example, in glycolysis a reaction that takes place in the cell cytoplasm, a glucose molecule is converted into two molecules of pyruvic acid by number of steps. If pyruvic acid is no more required then the very first step of the reaction *i.e.*, conversion of glucose to glucose 6- phosphate is stopped. This reaction is catalyzed by the enzyme hexokinase. The activity of this enzyme is inhibited by glucose 6 phosphate as shown in the reaction below.

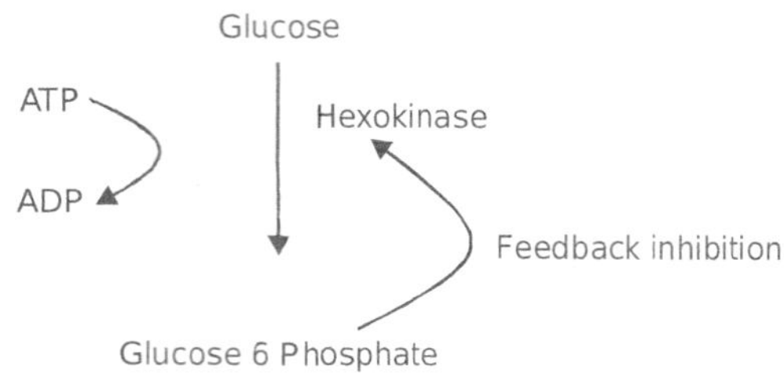

Fig. 3.6.

This type of inhibition is also known as allosteric modulation. In allosteric modulation inhibitor attaches with the enzyme at some site other than the active site and changes the conformation of enzyme and stops the reaction.

COENZYMES

Coenzyme is an organic compound which acts as donor or acceptor of electrons, specific atoms or groups that are added or removed from the substrate. Many coenzymes are associated with enzyme catalyzing oxidation – reduction reaction *e.g.*, nicotinamide adenine dinucleotide (NAD) which is a carrier of electrons and hydrogen is associated with the enzyme alcohol dehydrogenase and helps in the conversion of acetaldehyde into alcohol, a reaction of alcoholic fermentation. It is also pertinent to mention here that one type of coenzyme is not only associated with the functioning of one enzyme but a number of different enzymes *e.g.*, NAD act as a coenzyme for many enzymes of cellular respiration. Some other coenzymes which are carriers of electrons and hydrogen are nicotinamide adenine dinucleotide phosphate (NADP), flavin mononucleotide (FMN), flavin dinucleotide (FAD) and coenzyme Q (CoQ). Another coenzyme (CoA) is a carrier of acyl group. Many coenzymes are closely related to enzymes *e.g.*, thiamine pyrophosphate (TPP) transfers an aldehyde group contains vitamin B_1. Similarly FMN contains riboflavin (vitamin B_2). Biotin is a vitamin and acts as carrier of CO_2. Unlike enzymes coenzymes are altered during the enzymatic reaction. Their original structure is restored by some other reaction.

COFACTORS

Certain metals present in trace amounts also facilitate enzyme reaction. Fe^{2+} is a cofactor responsible for the catalytic action of catalase. Similarly Zn^{2+} ions are cofactors of the enzyme of carbonic anhydrase as well as alcohol dehydrogenase and molybdenum is a cofactor of enzyme nitrate reductase. Mg^{2+} is a cofactor of kinase type of enzyme. Other metal ions which act as cofactors are Cu^{2+}, Cu^{+} and K^{+}.

Exercises

1. What are enzymes ? Differentiate between enzymes and inorganic catalyst.
2. Describe the characteristics of enzymes.
3. Explain Lock and Key hypothesis of enzyme action.
4. Describe the mechanism of enzyme action.
5. Give a brief account of various factors affecting enzyme activity.
6. How enzyme activity is inhibited ? Describe by giving suitable examples.
7. Write a brief note on coenzymes and cofactors.

❑❑❑

UNIT 2

- ★ Plant System
- ★ Animal System
- ★ Microbial System

Chapter 4 PLANT SYSTEM

Plants are the autotrophic organisms. They can synthesize their own food from carbon dioxide and water in the presence of sunlight with the help of chlorophyll and photosynthetic pigments. There are about 400,000 different kinds of plants. They occur both in aquatic and terrestrial habitats. They vary in shape and size. Some of the plants are microscopic and can be seen with the help of a microscope whereas many type of trees are reaching over 100 m in height. There are two main groups of plants: cryptogams and phanerogams.

CRYPTOGAMS

Crytogams (*kryptose*-concealed; *gamos*-marriage) are the flowerless or seedless plants and phanerogams. Cryptogams are further divided into three groups: Thallophyta,Bryophyta and Pteridophyta.Thallophytes are the simplest plants like algae and fungi. Their body is like a thallus *i.e.,* not differentiated into root, stem and leaf. Examples of algae are *Spirogyra*, *Chara*, fresh water forms and *Ectocarpus* and *Polysiphonia*, marine algae. Fungi are the other thallophytes which unlike other plants are heterotrophic in their mode of nutrition. Some of them are saprophytes growing on dead and decaying organic matter *e.g., Penicillium, Mucor* and others are parasites *e.g., Puccinia* on wheat and *Phytophthora* on potato. Bryophytes are known as amphibians of plant kingdom and their body is either thalloid or leafy axis a few cm. in size which are found in wet and shady places, *e.g., Marchantia* and *Funaria*. Pteridophytes are known as vascular cryptogams as they posses xylem and phloem tissues for the conduction of water and photosynthates respectively. Important examples of pteridophytes are ferns which grow in hills as well as in plains in moist places.

PHANEROGAMS

Phanerogams (*phaneros*-visible; *gamos*-marriage) are the flowering or seed bearing plants. They are further divided into two groups: Gymnosperms and Angiosperms. Gymnosperms are the naked seeded plants which are mostly trees and a few are shrubs. Important examples of this group are *Cycas* and *Pinus.* Angiosperms are the closed seeded plants and at present constitute the dominant vegetation. They are of two sub types dicots and monocots. Dicots have two cotyledons in their seeds whereas monocots have a single cotyledon in their seeds. Examples of dicots are pulses and tree like mango and banyan. Cereals like wheat and rice are some of the monocots.

PARTS OF AN ANGIOSPERMIC PLANT

Body of a typical angiospermic plant is consists of root, stem, leaf, flower and fruit (Fig. 4.1). Each part performs specific functions. Root is the underground part of the plant which is positively geotropic that fixes the plant into the soil and absorbs water and nutrients from the soil. Stem is the above ground part of the plant that is negatively geotropic and it supports other plant parts like branches, leaves, flowers and fruits. It transports water and mineral nutrients from the roots to the other body parts. Through it sugars are transported from the leaves to the non green parts like roots. Leaves perform the function of photosynthesis and the exchange of the respiratory gases. In this process glucose, a carbohydrate is synthesized from water and carbon dioxide with the help of chlorophyll in the presence of sunlight and oxygen gas is released. Function of flower is in reproduction. It consists of stamens and carpels which are the male and female reproductive parts of the flower respectively and are the essential parts of a flower. Sepals and petals are the accessory parts of a flower. After fertilization ovary of the carpel matures into a fruit that contains one or many seeds. A seed is a mature and ripened ovule. Seeds germinate to give new plants. Plant is an autotrophic system as shown in Fig. 4.2. In this chapter some important plant physiological processes like nutrition, nitrogen metabolism, photosynthesis and growth have been explained.

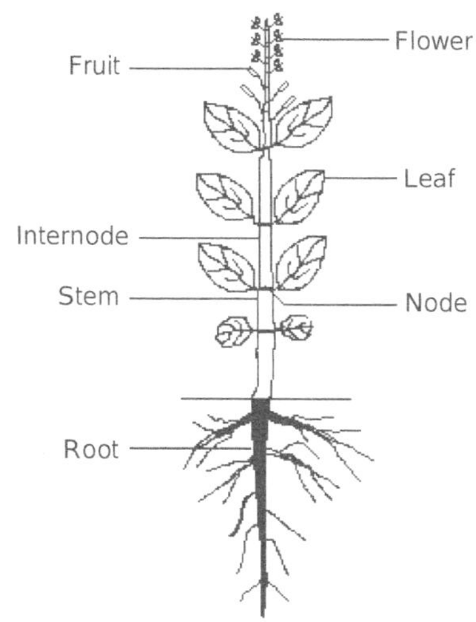

Fig. 4.1. Parts of an angiospermic plant.

MINERAL NUTRITION IN PLANTS

Plants require a number of mineral elements which are utilized by them for proper growth and development. Chemical analysis of plants reveals that they contain nearly 40 elements of the periodic table. Out of which only a few are required by the plants. **Sachs** and **Knop** (1850) demonstrated that carbon (C), hydrogen (H), oxygen (O), nitrogen (N), phosphorous (P), potassium (K), calcium (Ca), sulphur (S), magnesium (Mg) and iron (Fe) are the elements which are necessary for the growth of plants. These 10 elements are required in more amounts and are known as macronutrients. Later on a few other elements like manganese (Mn), zinc (Zn), boron (B), copper (Cu) and molybdenum (Mo) were also found to be essential for plants. These 5 elements are required in extremely small amounts and are called micronutrients. Criteria for essentiality of mineral elements were given by **Arnon** and **Stout** (1939). A mineral element is essential for plants if it fulfills the following criteria.

(*i*) A plant cannot complete its life cycle in the absence of an essential element.

(*ii*) An essential element cannot be replaced by any other element.

(*iii*) An essential element must be involved directly in the nutrition of a plant.

Availability of Essential Elements

As listed earlier there are 10 macronutrients and 5 micronutrients that are essential for the normal growth and development of majority of plants. Elements like C, H and O are not the mineral elements. These are obtained from air and water. C and O are obtained in the form of gases like CO_2 and O_2 which constitute 0.032% and 21% respectively of the air. H is obtained from water (by photolysis during photosynthesis) and water is absorbed by plant roots from the soil. Plants also liberate O_2 during photosynthesis and CO_2 during respiration. Rest of the elements are obtained from the soil which is the store house of all essential mineral elements. These elements are not present in their free state but in combined state as oxides, chlorides, sulphates, etc. in the soil and are usually absorbed in ionic form by plant roots. There are two methods of absorption : passive uptake and active uptake.

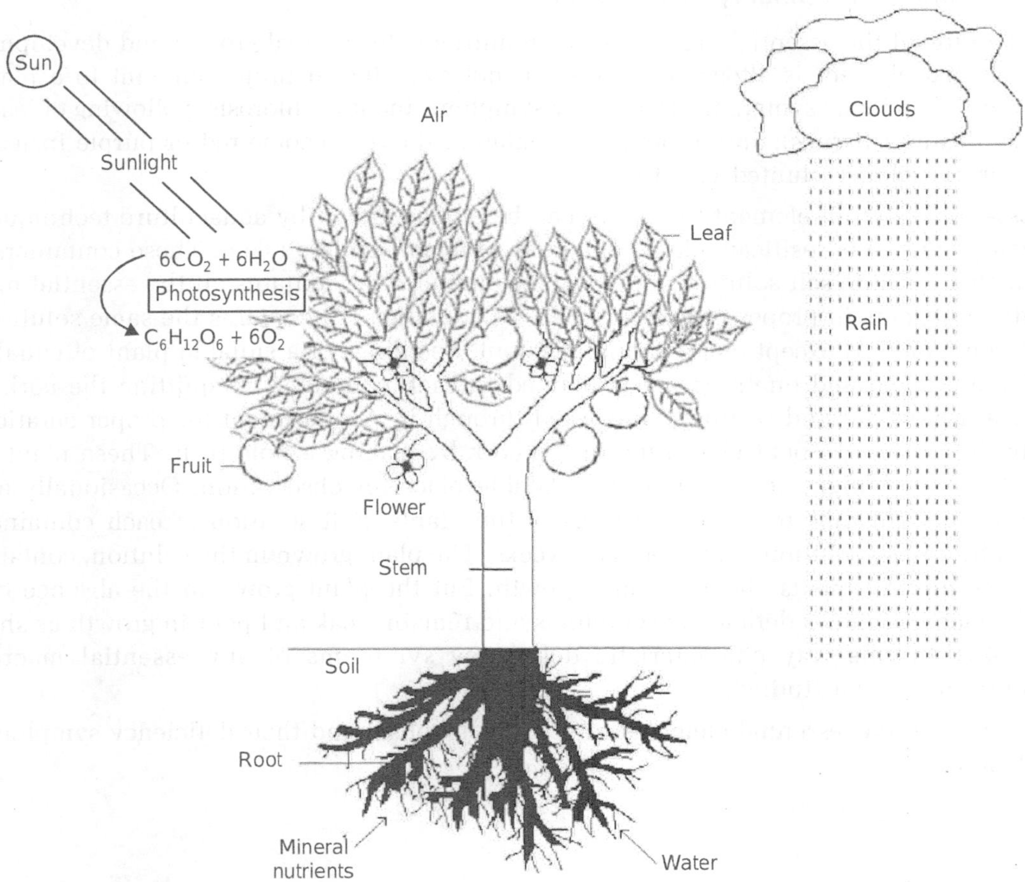

Fig. 4.2. Plant body – An autotrophic system.

Passive Absorption

In this method no expenditure of energy takes place and minerals are absorbed by diffusion. Diffusion is a physical process in which solutes move from a region of higher concentration to a region of lower concentration. Amount of minerals is more in soil as compared to plants so they are absorbed by diffusion inside the plant.

Active Absorption

In this method plants utilize energy to absorb minerals. It is necessitated as some minerals are required more in amount as compared to their presence in the soil. There are certain carrier molecules made up of proteins present in the plasma membrane of plant cells. These carrier molecules when energized by ATP molecules help in transport and accumulation of salts inside the cell. *Nitella*, a green alga, can accumulate 25 times more potassium inside its cell as compared to the concentration of potassium in the water in which it grows.

Demonstration of Essentiality of Elements

Plants require all the essential macro and micronutrients for normal growth and development. If an essential element is altogether lacking or not available in proper amount to a plant it shows some deficiency symptoms. Deficiency symptoms include chlorosis (yellowing of leaves), necrosis (death of tissues), anthocyanin pigmentation (leaves become red or purple instead of normal green colour), stunted growth, etc.

Essentiality of an element to a plant can be demonstrated by aquaculture technique. In this technique two borosilicate glass containers of equal size are taken. These containers are filled with 1% dilute salt solution. One of the containers (A) contains all the essential macro and micronutrients in proper proportion. Second container (B) contains the same solution as that of container (A) except one essential element. Seedlings of a suitable plant of equal age and size are taken and one seedling is planted in each container by splitting the cork and making a narrow slit and seedling is inserted through it. Arrangement for proper aeration is done by introducing a bent glass tube in the cork by making a hole in it. These plants are exposed to the sunlight and are put at a suitable place for observation. Occasionally air is blown in the bent tube to aerate the roots of the plants. Salt solution of each container is replaced by a fresh solution after every two weeks. The plant grown in the solution, containing all the essential elements shows normal growth. But the plant grown in the absence of an essential element shows deficiency symptoms and remain weak and poor in growth as shown in Fig. 4.3. In this way characteristic deficiency symptoms of any essential macro or micronutrient can be studied.

Role of various essential elements in plant metabolism and their deficiency symptoms is given below:

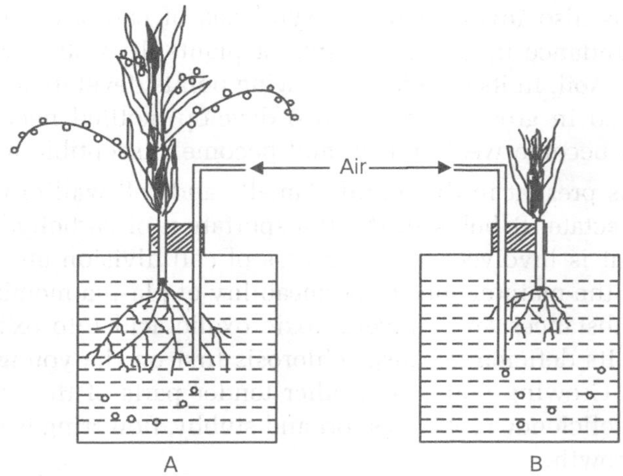

A. Plant grown in a solution containing all essential mineral nutrients
B. Plant grown in the absence of an essential mineral nutrients
Fig. 4.3. Aquaculture technique.

Macronutrients

All the macronutrients except potassium are the constituents of different biomolecules which are the structural components of the cells and are required in performing different functions. Potassium is essential in carrying out some other functions. Among the macronutrients C, O and H constitutes about 92% of the plant body and are not mineral in origin. They are obtained in some other form as discussed in the availability of essential elements to the plants. Macronutrients like N, P, K, Ca, Mg, S and Fe are mineral in origin obtained from soil and together constitute about 5–6% of body mass of plants in which major share is that of nitrogen.

(*i*) ***Nitrogen.*** It is the most important macronutrient and is present in proteins, enzymes, purines, pyrimidines, vitamins etc. Plants absorb nitrogen chiefly in the form of nitrates and ammonium ions from the soil. Its deficiency causes chlorosis due to low content of chlorophyll. Chlorosis is first noticed in the younger leaves then in the mature leaves. Nitrogen is a mobile element, younger leaves retain their nitrogen and also get it from old and mature leaves. Sometimes, leaves develop anthocyanin pigmentation as they turn red or purple by production of pigments other than chlorophyll. Excess of nitrogen results in vigorous vegetative growth, leaves become dark green and flowering is delayed.

(*ii*) ***Phosphorous.*** It is a structural component of ATP, nucleotides of DNA and RNA, coenzymes like NAD, NADP and TPP. It is also present in phospholipids which are found in biological membranes. It is absorbed chiefly as phosphate ions. Its deficiency affects the overall metabolism of plants. In its deficiency plants remain stunted in growth and developed necrotic areas on leaves and fruits. Purple anthocyanin pigmentation also takes place in leaves due to its deficiency. Deficiency symptoms first appear in old leaves then in younger leaves.

(*iii*) ***Potassium.*** It is the only macronutrient which is not a structural component of any organic molecule of the cell but play a vital role in plants to carry out functions like opening and closing of stomata and as a cofactor (activator) of a number of respiratory enzymes. It is also involved in the synthesis of carbohydrates and proteins. It is found in abundance in growing points of plants. It is absorbed chiefly in its ionic form from the soil. In its deficiency growing points develop necrotic areas and plants remain stunted in growth. Leaves also develop mottled necrosis. In its deficiency cereal plants become weak, bushy and become susceptible to lodging.

(*iv*) ***Calcium.*** It is present in the middle lamella and cell wall of plant cells in the form of calcium pectate. It helps in the transportation of carbohydrates and in the lipid metabolism. It is involved in the process of cell division and cell enlargement and also helps in the maintenance of permeability of plasma membrane. It is antitoxic to poisonous substances and converts toxic oxalic acid into oxalate. It also promotes root growth. Its deficiency causes chlorosis first on the younger leaves then on the older leaves. Growing points and other tender parts of the plants develop necrotic spots. In its deficiency roots are short and stubby, flowering is less and plants remain stunted in growth.

(*v*) ***Sulphur.*** It is present in proteins as sulphur containing amino acids cysteine and methionine. It is found in coenzyme A and vitamins like thiamine and biotin and thus very important for plant metabolism. It helps in nodule formation in the leguminous plants. The characteristic smell of mustard oil is due to sulphur containing volatile compound *sinigrin*. It is absorbed as sulphate ions and its deficiency causes chlorosis first on the younger leaves then on the older ones.

(*vi*) ***Magnesium.*** It is a constituent of chlorophyll, a green pigment, present in leaves and is necessary for the process of photosynthesis. It also acts as an activator of several enzymes of cellular respiration. It is essential for binding of component of ribosomes during the process of biosynthesis of protein. Its deficiency causes interveinal chlorosis, anthocyanin pigmentation and necrosis in leaves.

(*vii*) ***Iron.*** As compared to other macronutrients iron is required in lesser amount by a plant. It is present in cytochromes and helps in the electron transport system of respiration and photosynthesis. It is also essential for chlorophyll synthesis. Its deficiency causes interveinal chlorosis, the symptom first comes on younger leaves then on older leaves.

Micronutrients

Micronutrients are required in very trace amounts usually in parts per million (ppm). Micronutrients like Mn, Zn, Cu, and Mo act as activators of different enzymes. B is the only micronutrient which does not acts as an activator of any enzyme but is required in some other metabolic activity of plants.

(*i*) ***Manganese.*** It acts as an activator of some respiratory enzymes required in dehydrogenation and decarboxylation reactions. It helps in the photolysis of water and release of O_2 during photosynthesis. Its deficiency causes necrosis and chlorosis in interveinal regions of leaves.

(*ii*) ***Zinc.*** It is an activator of enzyme alcohol dehydrogenase. It helps in the synthesis of a phytohormone, auxin, indole acetic acid (IAA) that regulates the growth of plants. Its deficiency retards plant growth, leaves develop chlorosis and plants develop characteristic *'little leaf'* and mottled leaf condition. In its deficiency internodes are shorten and the herbaceous plants acquire rosette type of growth.

(*iii*) ***Copper.*** It is a component of plastocyanin, a compound present in chloroplast and is required in electron transport system of photosynthesis in plants. It acts as a cofactor for enzymes like polyphenol oxidase and plays an important role in plant metabolism. Its deficiency develops 'die back' symptoms which are characterized by the appearance of necrotic spots on leaves tips and spreading along the leaves margins. Ultimately the leaves wither away and plants give a wilting appearance. In fruit trees deficiency of copper causes a disease 'exanthema' in which irregular brownish spots with gummy exudates appear on the leaves and fruits.

(*iv*) ***Molybdenum.*** It acts as an activator of the enzyme nitrate reductase and plays an important role in plant metabolism. Its deficiency causes necrosis and chlorosis in leaves. Deficiency symptoms first appear on older leaves then in younger ones.

(*v*) ***Boron.*** It helps in the absorption of calcium and translocation of carbohydrates. It is also involved in the process of self differentiation and germination of pollen grains. In its absence growth is retarded and plants bear fewer flowers. Its deficiency also results in disintegration of internal plant tissues that causes 'heart rot' of beets and 'drought rot' of apples.

NITROGEN METABOLISM IN PLANTS

Nitrogen is a critical element not only for plants but for all other living organisms. It ranks fourth after carbon, oxygen and hydrogen in the composition of plants. It is required in the synthesis of many cellular components of plants including DNA, chlorophyll, enzymes and proteins. Atmospheric air is the main reservoir of nitrogen. It is present in the gaseous state as dinitrogen (N_2) and constitutes nearly 78% of the air. Nitrogen is an inert gas and as such is of no use to the plant. It is similar to a situation in which plants are submerged in a 'sea of nitrogen' still they cannot make any use of it. Plants can absorb nitrogen from the soil in the form of some fixed and soluble ionic compounds like; nitrate (NO_3^-), nitrite (NO_2^-) and ammonium (NH_4^+) ions and this is made possible only by some nitrogen fixing bacteria.

Nitrogen Fixation

It is a process by which nitrogen gas present in the environment is converted into ammonia and that is added into the soil by some nitrogen fixing prokaryotic microorganisms like bacteria and cyanobacteria. There are two types of nitrogen fixing bacteria symbiotic (*Rhizobium* and *Frankia*) and free living (*Azotobacter* and *Clostridium*). *Rhizobium* is found in the root nodules of leguminous plants like pea, gram, ground nut, etc. *Frankia* is a filamentous bacterium (actinomycetes) that is found in the root nodules of trees like *Ulnus* and *Casuarina*. *Azotobacter* and *Clostridium* are saprophytic nitrogen fixing bacteria found in the soil. Some of the blue green algae (cyanobacteria) *Nostoc* and *Anabaena* found in aquatic habitat are also nitrogen fixing organisms. *Nostoc*, besides growing freely in water also occurs as a symbiont in *Anthoceros* (a bryophyte), *Azolla* (a water fern) and roots of *Cycas* (a gymnosperm).

Among the above mentioned nitrogen fixing microorganisms the most important is the *Rhizobium*. It occurs in the soil as a saprophyte but when come in contact with roots of a leguminous plant it enters into it, multiply there to form a number of bacteriods that grow in the cells of root cortex forming root nodules. Root nodules are small swellings present on the root surface as shown in Fig. 4.4.

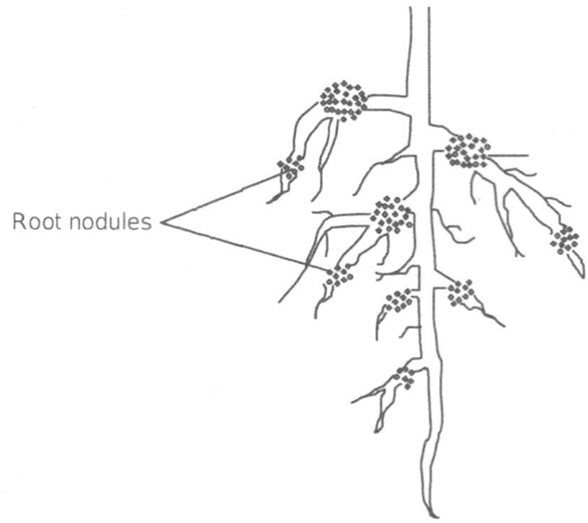

Fig. 4.4. Nodulated root of a leguminous plant.

Rhizobia form root nodules only in leguminous plants not in other plants as there is compatibility between lectins (a type of protein) present in legumes and the polysaccharides of the capsule of *Rhizobia*. There is a symbiotic relationship between *Rhizobia* and legumes in which both the organisms are mutually benefitted. *Rhizobium* gets sugars from the leguminous plant and in turn the later provides nitrogen rich compounds ammonia (NH_3) and amino acids to the legumes. *Rhizobium* can convert nitrogen gas into ammonia only when it is present inside the root nodules with the help of nitrogen fixing enzyme nitrogenase. It is a complex enzyme consisting of two components a molybdenum – protein complex and an iron – protein complex. These two components work in close association to bring about reduction of nitrogen into ammonia.

$$N_2 + 6H^+ + 6e^- + 12ATP \xrightarrow{\text{Nitrogenase}} 2NH_3 + 12ADP + 12iP$$

The enzyme nitrogenase requires an oxygen free environment that is regulated by leghaemoglobin a pigment similar to blood haemoglobin present in root nodules. Leghaemoglobin is synthesized by the interaction between *Rhizobia* and root cortex of legumes. Nitrogenase also requires energy that is supplied by ATP and a reducing agent. Reducing agent in this reaction that supplies hydrogen (electrons and protons) is either a reduced ferredoxin or a flavodoxin. These compounds are reduced with the help of NADH or NADPH by the metabolism of leaf derived glucose in the nodules.

Nitrification

Although plants can absorb ammonia from the soil but it is rapidly converted into nitrites and nitrates by chemosynthetic soil bacteria *Nitrosomonas* and *Nitrobacter* respectively.

$$\text{Ammonia} \xrightarrow{\text{Nitrosomonas}} \text{Nitrite} \xrightarrow{\text{Nitrobacter}} \text{Nitrate}$$

Chemosynthetic bacteria are a group of autotrophic bacteria which synthesize sugars with the help of energy derived from oxidation of some inorganic compounds. Some nitrates are also formed by electrification during thunder storms as atmospheric gases N_2 and O_2 reacts to form oxides of nitrogen which come to the land along with rain water. Some ammonia and nitrates are lost from the soil by denitrifying bacteria like *Pseudomonas*. These bacteria convert back ammonia and nitrates into nitrogen gas and this process is termed as denitrification.

Assimilation of Nitrogen

Nitrates absorbed by the plants from the soil are again converted into ammonia inside the plant body for its utilization. Nitrates are first converted into nitrites with the help of an enzyme nitrate reductase and nitrites are further converted into ammonia by the enzyme nitrite reductase. The enzyme nitrate reductase requires a reducing agent *i.e.*, reduced flavodoxin ($FADH_2$) which is itself reduced either by NADH or NADPH. Molybdenum acts as a cofactor of the enzyme nitrate reductase that transfers electrons from the reducing agent to reduce nitrate into nitrite. This conversion takes place both in roots and leaves.

$$\text{Nitrates} \xrightarrow{\text{Nitrate reductase}} \text{Nitrites} \xrightarrow{\text{Nitrite reductase}} \text{Ammonia}$$

Reduction of nitrite into ammonia takes place chiefly in the leaves and reduced ferredoxin is the reducing agent in this conversion.

Ammonia so formed is incorporated into keto-acids to synthesize amino acids which are considered as initial products of nitrogen assimilation in plants. Keto-acids are synthesized during the Kreb's cycle in cellular respiration that acts as a carbon skeleton of amino acids. Amino acids are synthesized by the following two methods:

(*i*) ***Reductive ammination.*** In this process ammonia reacts with α – keto glutaric acid in the presence of an enzyme glutamate dehydrogenase to synthesize glutamic acid (an amino acid).

$$NH_3 + \alpha\text{-keto glutaric acid} \xrightarrow[\text{NADH} \rightarrow \text{NAD}^+]{\text{Glutamate dehydrogenase}} \text{Glutamic acid}$$

(*ii*) ***Trans ammination.*** Once glutamic acid is synthesized as many as 17 different amino acids can be synthesized from it by trans ammination reactions. In a trans ammination reaction amino group of glutamic acid is transferred to another keto acid which gets transformed into an amino acid as shown below:

$$\text{Glutamic acid + Oxalo acetic acid} \xrightarrow{\text{Glutamic-aspartic transaminase}} \text{Aspartic acid} + \alpha\text{-keto glutaric acid}$$

PHOTOSYNTHESIS

It is a process by which plants manufacture their own food in the form of sugars (organic compounds) from simple inorganic molecules like water and carbon dioxide in the presence of sunlight with the help of green pigment chlorophyll. In this process solar energy is converted into chemical energy. Photosynthesis occurs in the green parts of the plants chiefly in their leaves. Plant require continuous supply of raw materials *i.e.*, water and carbon dioxide to carry out photosynthesis. Plants absorb water and minerals nutrients by their roots from the soil and are transported to other parts through the stem with the help of a complex tissue xylem. Plants obtain CO_2 from air through stomata (minute apertures) present on the leaf surface. CO_2 is also released by the plants during respiration. Over all reaction of photosynthesis is represented as:

$$6CO_2 + 12H_2O \xrightarrow{\text{Chlorophyll}} \underset{\text{Glucose}}{C_6H_{12}O_6} + 6H_2O + 6O_2$$

Photosynthesis reaction is appropriately called as *'Reaction of life'* because plants are autotrophs and other living organisms derive their food directly or indirectly from the plants. Oxygen gas (O_2) released in this process is also used by other organisms.

Chloroplast

Chloroplast is a semiautonomous cell organelle present only in plant cells. Chloroplasts containing cells are found in the green parts of the plants chiefly in the mesophyll cells of the leaves. In each of the mesophyll cell there are about 300 chloroplasts. There are about 300 chloroplasts in each of the mesophyll cell. Chloroplast is generally oval, ellipsoidal or lens shaped structure which is about 5 μm – 10 μm in diameter and 2 μm – 3 μm in thickness. It is bounded by double membranes an outer and an inner. Internally chloroplast contains two components: thylakoids (membranous lamellae) and a proteinaceous stroma in its matrix as shown in Fig. 4.5. Thylakoids contain photosynthetic pigments; chlorophylls and carotenes. Chloroplast is the cell organelle in which photosynthesis occurs which is completed in two successive steps known as light reaction and dark reaction. Light reaction of photosynthesis takes place in the thylakoids and the site of the dark reaction is the stroma of chloroplast.

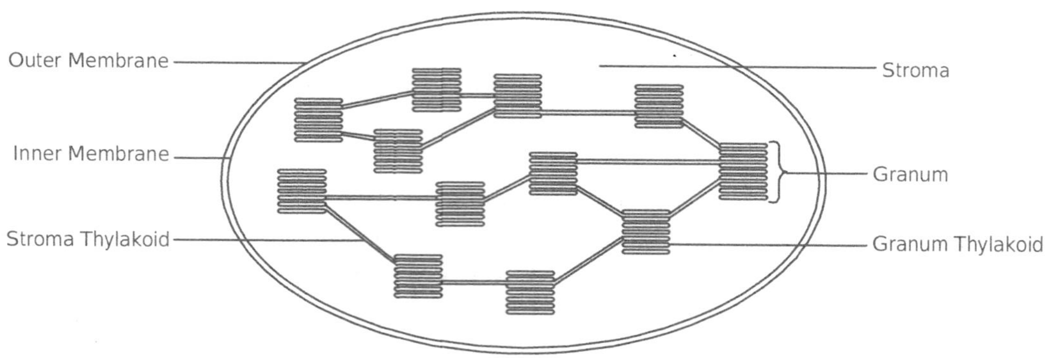

Fig. 4.5. Stucture of chloroplast.

Thylakoids are arranged like the stacks of the coins to constitute grana. There are about 10–90 thylakoids in each granum. There are about 40–60 grana in a chloroplast. Grana are linked by stroma thylakoid. Photosynthesis is completed in two phases: light reaction and dark reaction occurring in thylakoids and stroma respectively are components of chloroplast.

Light Reaction

This reaction takes place in thylakoids in the presence of sunlight. In this reaction different pigments present in thylakoids harness the sunlight to form 'assimilating power' *i.e.,* ATP an energy rich compound and NADPH a reduced coenzyme. To trap sunlight, thylakoids contains two types of pigments: chlorophylls and carotenoids. Chlorophylls are of two type chlorophyll – *a* (green) and chlorophyll – *b* (bluish green). Carotene (yellow) and xanthophylls (pink) are two carotenoids. Chlorophylls absorb sunlight in blue and red region of visible spectrum of sunlight (400 nm – 700 nm). Carotenoids absorb light mainly in violet-blue region. Chlorophyll-*a* is the principal pigment required in light reaction and is found in abundance in the leaves. It is a complex molecule consisting of two regions: a head and a tail. Head region consists of centrally placed Magnesium atom (Mg) surrounded by a porphyrin ring. This ring is linked to Mg atom by its four nitrogen atoms by co-ordinate bonds. Tail part is long consisting of hydrocarbons and is known as phytol. Chlorophyll–*a* and *b* resembles in their structure and the only difference is with respect to the position marked as R of porphyrin ring as shown in Fig. 4.6. At position R Chlorophyll–*a* contains a methyl group ($-CH_3$) and Chlorophyll-*b* has an aldehyde group (–CHO).

Photosynthetic pigments absorb variable amount of light incident on the leaves. Absorption spectrum of chlorophyll-*a* which is the principal photosynthetic pigment is shown in Fig 4.7. As depicted from the absorption spectrum of chlorophyll-*a*; it absorbs sunlight mainly in the blue and red regions and reflects light in the green region; it is for this region that the leaves look green. Absorption sunlight is harnessed by chlorophylls and other photosynthetic pigments initiate light reaction.

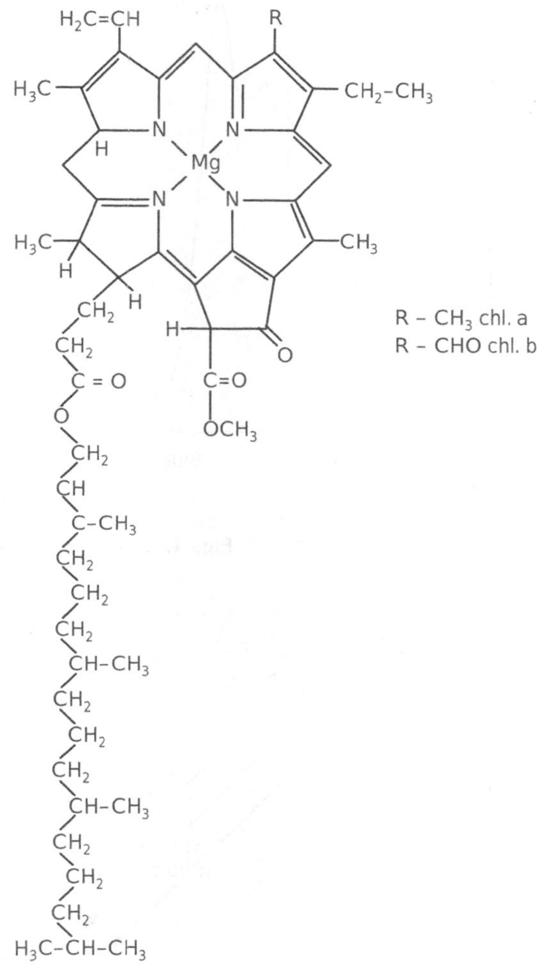

Fig. 4.6. Structure of chlorophylls.

Light Harnessing Unit

Different photosynthetic pigments are embeded in the thylakoids membrane and are present in groups of about 200 to 250 molecules and constitute a light harnessing unit known as photosystem. Each such unit is composed of a special type of chlorophyll–*a* molecule known as reaction center and is surrounded by hundreds of other chlorophylls and carotenoids pigments that are known as accessory or antenna pigments. When chlorophyll–*a* molecule of a reaction center absorbs sufficient light its electron gets excited to a higher energy state. Chlorophyll–*a* itself cannot absorb the required light to excite its electrons, the antenna pigments help in capturing the sufficient quantum of light to excite the electrons of the reaction center as shown in Fig. 4.8. There are two types of photosystems: Photosystem I (PS–I) and Photosystem II (PS–II). Reaction center of PS–1 and PS–II is composed of special types of chlorophyll molecules P–680 and P–700 respectively.

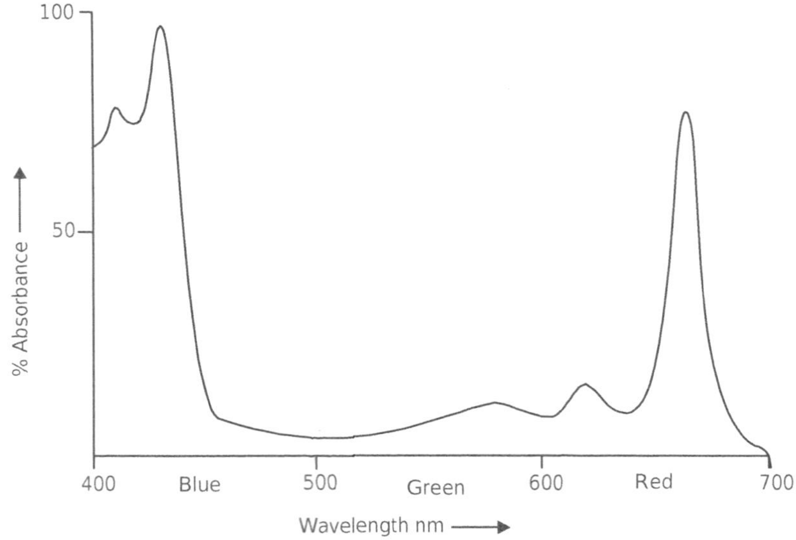

Fig. 4.7. Absorption spectrum of chlorophyll-*a*.

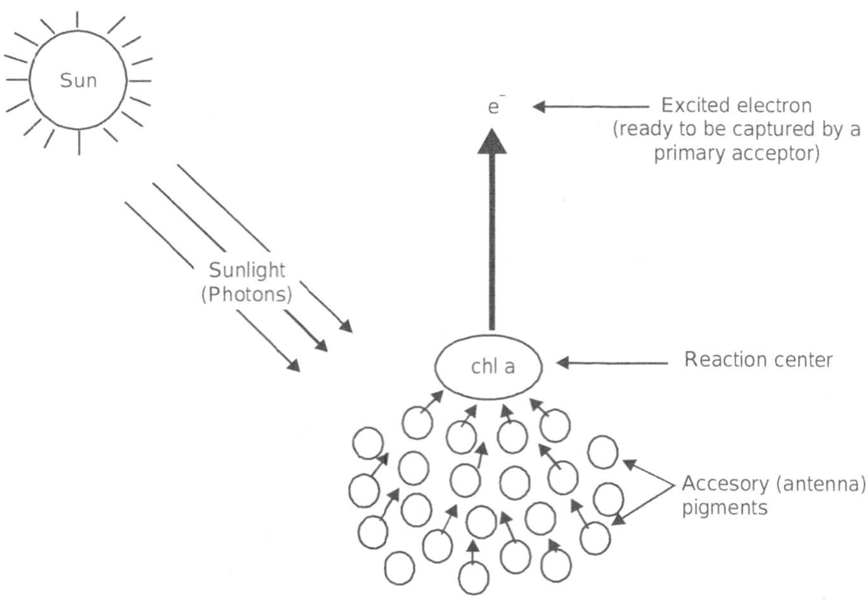

Fig. 4.8. Harnessing of sunlight by photosynthetic pigments.

Mechanism of Light Reaction

Light reaction is fundamentally an oxidation – reduction process occurring in chloroplast by the combined action of both photsystem-I and photosytem-II as shown in Fig. 4.9. Excited electrons from the reaction center of photosystems are accepted by $NADP^+$ which gets reduced as NADPH + H^+. Photosystems are activated in the sunlight. PS–II is activated at or below 680 nm of wavelength and PS–I is activated at a wavelength of 700 nm or below. Excited electrons of PS–II are accepted by Pheophytin which is the primary acceptor of electrons in this case. The oxidized chlorophyll a molecule of PS–II brings about photolysis of water to regain its electrons.

$$H_2O \longrightarrow 2e^- + 2H^+ + 1/2\ O_2$$

Electrons from pheophytin are transferred to PS–I by electron carriers like plastoquinone (PQ), cytochromes complex b_6–f and plastocyanin. Excited electrons from PS–I are accepted by Fe – S protein which is the primary acceptor of electrons in this case. Thus, the electrons lost by the PS–I are regained by the electrons of PS–II. Ferredoxin accepts electrons from Fe – S proteins that reduce NADP+. The enzyme NADP reductase catalyzes this reaction. In light reaction energy rich ATP molecules are synthesized by the energy released during the down hill transfer of electrons as shown in Fig. 4.9. This is an old concept of ATP generation. According to chemosmotic hypothesis, which is a new concept, synthesis of ATP molecule takes place when protons which get accumulated in the lumen of thylakoids by the photolysis of water come out into stroma. Energy released in this process is utilized to generate ATP.

$$\text{ADP} + \text{iP} \xrightarrow{\text{ATP - synthase}} \text{ATP}$$

Enzyme ATP–synthase catalyze this reaction and this process is called photo-phosphorylation.

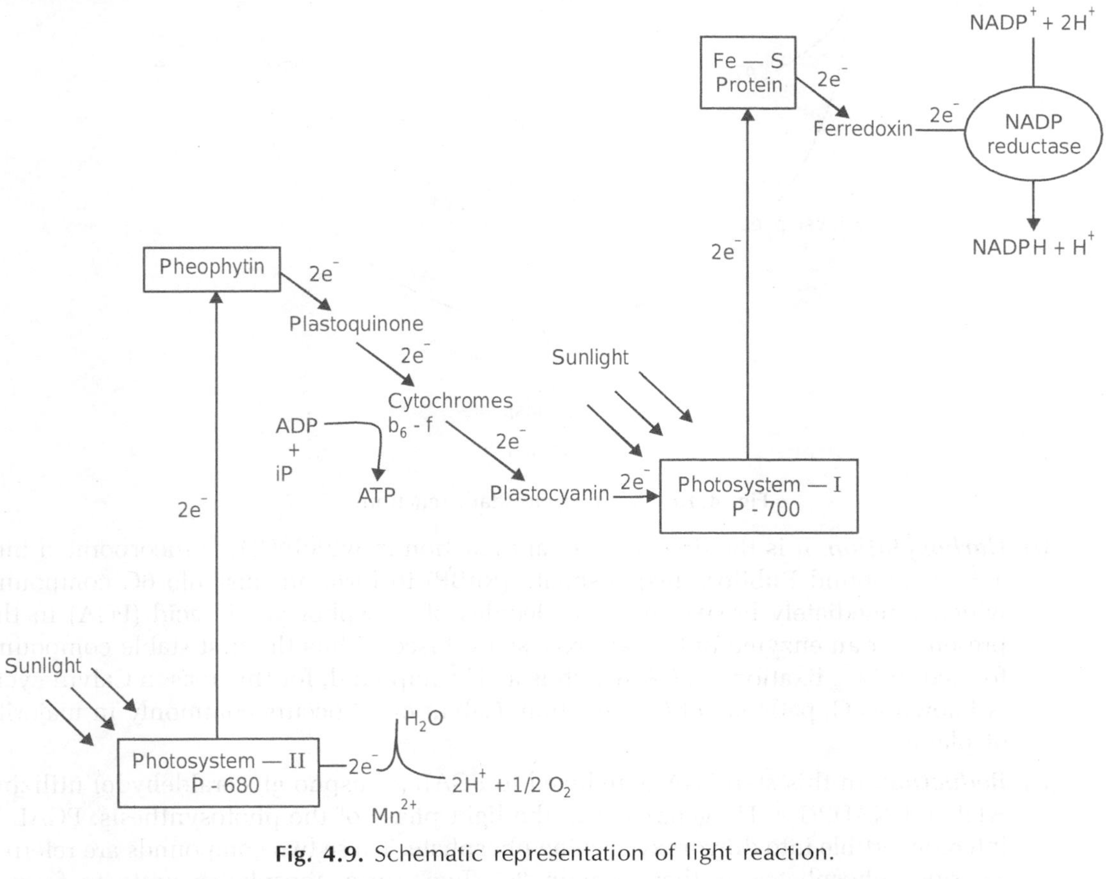

Fig. 4.9. Schematic representation of light reaction.

Oxygen is also released by PS–II during the light reaction along with the generation of 'assimilating power' (ATP and NADPH).

Dark Reaction

It takes place in the stroma of chloroplast and is independent of light and is also known as Calvin cycle. In this reaction CO_2 is reduced into sugars utilizing the assimilating power generated in the light reaction. Dark reaction was first elucidated by **Calvin, Benson** and co workers in 1954. It takes place in a cyclic manner and is also known as C_3 pathway of CO_2 fixation as it is completed in three phases as shown in Fig. 4.10.

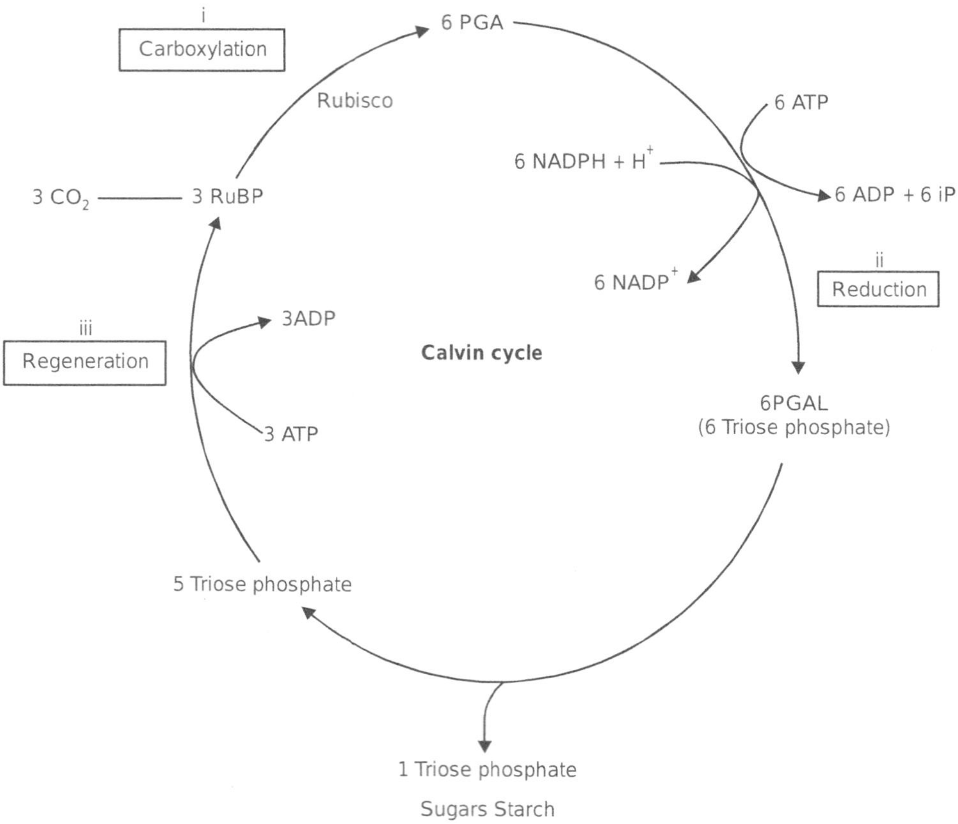

Fig. 4.10. Calvin cycle (dark reaction).

(*i*) ***Carboxylation.*** It is the first step of dark reaction in which CO_2 is incorporated into a 5C compound Rubilose bisphosphate (RuBP) to form an unstable 6C compound which immediately breaks into 2 molecules of phosphoglyceric acid (PGA) in the presence of an enzyme RuBP carboxylase (Rubisco). Thus the first stable compound formed in CO_2 fixation is PGA which is a 3C compound, for this reason Calvin cycle is known as C_3 pathway of CO_2 fixation. Calvin cycle occurs commonly in majority of plants.

(*ii*) ***Reduction.*** In this step PGA is reduced to PGAL (phospho glyceraldehyde) utilizing ATP and NADPH + H^+ generated in the light phase of the photosynthesis. PGAL is interconvertible into dihydroxy acetone phosphate. These two compounds are referred as triose phosphates as they contain 3C. Two triose phosphates unite to form a glucose molecule. Glucose is used as the main source of energy by the plants and excess of it is stored as starch which is a polymer of glucose in the chloroplast.

(*iii*) ***Regeneration.*** RuBP is regenerated from the triose phosphates by many inter conversion reactions to initiate dark reaction again. In this manner CO_2 is fixed in the form of carbohydrates.

Hatch and Slack Pathway (C_4 Pathway)

It is an alternative pathway of CO_2 fixation occurring in plants like sugar cane, maize etc. It was first elucidated by **Hatch** and **Slack** (1956). The first stable compound formed in Hatch and Slack pathway is a 4C compound oxalo acetic acid. Plants in which this pathway occurs are called C_4 plants which have a special type of leaf anatomy known as *'Kranz anatomy'* as shown in Fig. 4.11.

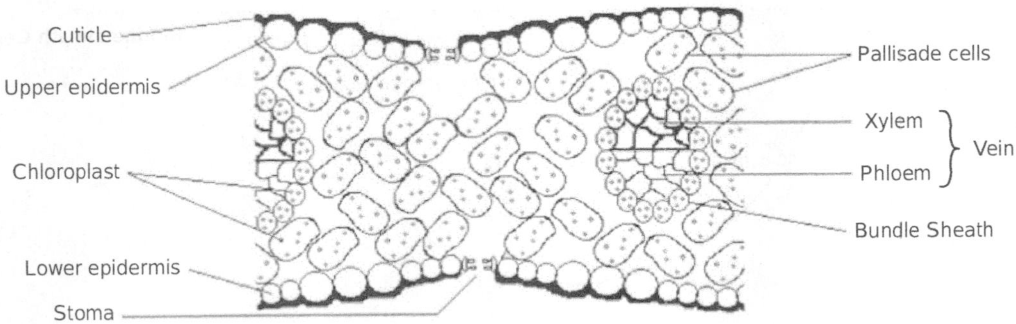

Fig. 4.11. T.S. of a C_4 leaf showing Kranz anatomy.

Leaf of such plants has two different types of chloroplasts in their mesophyll and bundle sheath cells. Mesophyll cells have predominance of the enzyme PEP–carboxylase and bundle sheath cells have the enzyme RuBP carboxylase. Bundle sheath cells are present around the veins of the leaf and contain chloroplast only in C_4 plant not in C_3 plants. Hatch and Slack pathway of CO_2 fixation is shown in Fig. 4.12. In C_4 plants CO_2 is incorporated in PEP (phosphoenol pyruvate) in the mesophyll cells to form OAA (oxalo acetic acid) in the presence of an enzyme PEP carboxylase. Oxalo acetic acid is reduced to malic acid. Malic acid may be aminated to form aspartic acid. Malic acid is transferred from mesophyll cell to the bundle sheath cell where it dissociates into pyruvic acid and CO_2 by an enzyme malic dehydrogenase. CO_2 enters into Calvin cycle and pyruvic acid is transported into the mesophyll cell where it is converted back into PEP utilizing energy. ATP is the energy source which is converted into AMP + iPP. This interconversion is equal to energy of 2 ATP molecules. PEP so formed again initiates the Hatch and Slack pathway.

C_4 plants are more efficient than C_3 plants in CO_2 fixation because at high O_2 concentration and high light intensity the enzyme Rubisco in C_3 plants performs oxygenase activity in which RuBP breaks into PGA and phosphoglycolate. PGA is utilized in the synthesis of sugars in the Calvin cycle but phosphoglycolate which is a 2C compound is converted immediately into glycolate. Glycolate metabolism in C_3 plants takes place with the help of 3 cell organelles namely chloroplasts, peroxisomes and mitochondria and is known as photorespiration. Glycolate is transported from chloroplast into peroxisome where it is converted into glycine. Then glycine is transported into mitochondria and is converted to serine and CO_2 without generation of ATP or NADH. C_3 and C_4 plants differ in many respects as shown in table 4.1

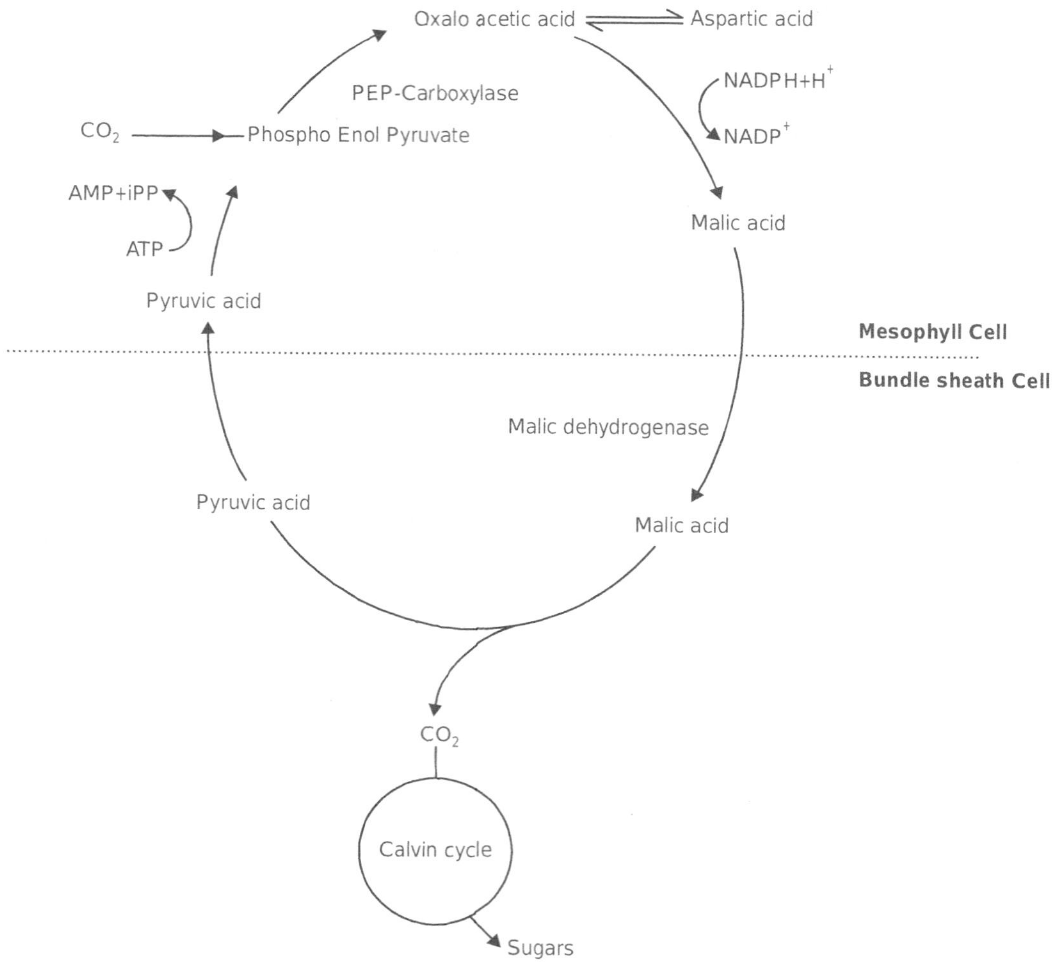

Fig. 4.12. Hatch and Slack pathway.

Table 4.1 Difference between C_3 and C_4 plants

	C_3 Plants	*C_4 Plants*
1.	CO_2 fixation in C_3 plants is by Calvin cycle and its initial acceptor is RuBP(rubilose bis phosphate)	CO_2 fixation in C_4 plants is by Hatch and Slack pathway and its initial acceptor is PEP (phosphoenol pyruvate).
2.	First stable compound in CO_2 fixation is a 3C compound PGA (phospho glycer-aldehyde).	First stable compound in CO_2 fixation is a 4C compound OAA (oxalo acetic acid).
3.	Do not have a Kranz type of leaf anatomy. In their bundle sheath cells there are no chloroplasts which are present only in mesophyll cells.	Leaf has a Kranz type of anatomy. Chloroplasts are present both in mesophyll and bundle sheath cells.

4.	Chloroplasts are of one type and present only in mesophyll cells and have the enzyme Rubisco.	Chloroplasts are of two types. Mesophyll cells have the enzyme PEP carboxylase in their chloroplasts and bundle sheath cells have the enzyme Rubisco.
5.	C_3 plants are less efficient in CO_2 fixation and some photosynthates are lost due to photorespiration.	C_4 plants are more efficient in CO_2 fixation as they check photorespiration.
6.	Food crops like wheat, rice, barley and potato are the important examples of C_3 plants.	Food crops like maize, sugarcane and millets are the important examples of C_4 plants.

Factors Affecting Photosynthesis

Process of photosynthesis is affected by a number of environmental factors like light, carbon dioxide, temperature and water. The effect of each factor is as follows:

(i) Light

It is the most important factor as in the process of photosynthesis light incident on the plants, is harnessed by the plant pigments to synthesize sugars from water and carbon dioxide; obliviously in its absence no photosynthesis takes place. It is the most detrimental factor as rate of photosynthesis is concerned and plants utilize about one percent of light incident on them and rest is reflected back in the atmosphere. Rate of photosynthesis is influenced by both quantity and quality of light. In low light intensity its rate is slow which increases as the intensity of light is increased. But at very high light intensity, photosynthesis rate drops down due to solarization; a phenomenon in which leaf cells are photooxidized and become non functional. Quality of light also influences the rate of photosynthesis. It has been observed experimentally that maximum photosynthesis occurs in red light, then in blue light and least in green light.

(ii) Carbon dioxide

Carbon dioxide present in the atmosphere constitutes a small fraction of the air, it is about 0.032 percent by volume of the total air and this much amount of it is sufficient for the sustainability of biological systems. It is this carbon dioxide which is reduced to sugars by the green plants with the help of solar energy; and the plants are a source of food for all the living organisms directly or indirectly. Carbon dioxide is released in the atmosphere by living organisms including plants during respiration. It has been experimentally observed that rate of photosynthesis is low at lower carbon dioxide concentration and the rate increases when its concentration is increased. However, at very high concentration of carbon dioxide, rate of photosynthesis falls as stomata get closed and entry of it into leaves from the air is stopped.

(iii) Temperature

During the process of photosynthesis a number of enzymes are involved and the activity of enzymes is influenced by temperature. The enzymes functions within an optimum range of temperature. In most of the plants rate of photosynthesis is increased as the temperature is increased from 0°C up to 40°C, and at a temperature above it the rate falls as plants loose much of water at high temperature due to transpiration and their stomata also get closed stopping the carbon dioxide intake.

(iv) ***Water***

Water is an essential raw material for photosynthesis. Though plants loose about 99 percent of water absorbed by them due to transpiration and retains only one percent of it that is utilized in photosynthesis. It has been observed that rate of photosynthesis is lower in water deficient plants. When less water is available to the plants, their stomata get closed restricting the entry of carbon dioxide in the leaves and thus lowering the rate of photosynthesis.

PLANT GROWTH

Growth is a complex process in which an irreversible increase in size and weight of a plant takes place by the cumulative effect of various metabolic activities. When seed of a plant germinates in soil under favourable environmental conditions its seedling emerges out. During seed germination its stored reserved food present in the form of carbohydrates, proteins or lipids is hydrolyzed by various hydrolytic enzymes and is utilized for the initial stages of growth. Inside the seed is present an embryo which is a miniature and dormant plant and is consists of two parts a plumule and a radicle. During seed germination radicle emerges out first which subsequently forms root system which is positively geotropic. Roots help in anchorage of plant into the soil and absorb water and minerals from it. Plumule emerges out a little later and forms the shoot system of the plant which is negatively geotropic. Shoot system is composed of stem, leaf, flower and fruit. Seedling depends on the stored reserved food only at the initial stages of growth later on it becomes independent as photosynthetic pigments are synthesized in its young leaf to carry out the process of photosynthesis. Photosynthetic activity in leaves and other green parts of the plant goes on adding photosynthates (sugars and starch) which help in increasing the number of cells, tissues and organs. A seedling is thus transformed into a young plant and later into a mature plant as shown in Fig. 4.13.

Fig. 4.13. Different stages of plant growth.
1-Seed, 2&3-Seed germination, 4-Seedling, 5-Young plant and 6-Mature plant

Growth in plants takes place by the meristematic cells (actively dividing cells) localized in special regions called meristems which may be apical, intercalary or lateral. Growth in length referred as primary growth takes place in all the seeded plants *i.e.*, monocots, dicots and gymnosperms. It is due to increase in number and enlargement of cells due to the activity of intercalary meristem (shoot apex and root apex). In some plants like, grasses increase in length is also due to the intercalary meristem. In dicots and gymnosperms increase in girth (thickness) that is referred as secondary growth is due to the meristematic activity of lateral meristem *i.e.*, cambium and cork cambium. Differentiation is also associated with the growth in which the new cells added by the meristems acquire specific shape and size on the basis of their location in the plant body to perform specific functions.

Requirements for Growth

Plants require water, minerals, oxygen, light and a suitable temperature for growth.

1. **Water.** It is an essential requirement of all the living organisms including plants. Bulk of plant body is composed of water. It is a major component of plant cells, provides turgidity to the cells and is a medium in which various metabolic activities takes place. Seed can only geminate under moist conditions. Plants absorb water with the help of roots from the soil. Though only about 1% of water absorbed by the plants is utilized in photosynthesis and other metabolic activities rest of it is transpired into the environment in the form of its vapours. In the absence of water, plants wilt and die under the prolonged drought condition.
2. **Mineral Nutrition.** Plants require different essential minerals to synthesize protoplasm. Minerals are made available to plants from the soil and are absorbed by their roots Minerals are essential for normal growth and development of plants.
3. **Oxygen.** Plants obtain O_2 from the air for respiration. O_2 is also released by the plants during photosynthesis. It is essential for oxidation of food to liberate energy and is utilized by the plants to carry out normal metabolic activities.
4. **Light.** Light is essential for plant growth. Plants synthesize their food by photosynthesis that occurs only in the presence of light. Light is also essential for flowering. Though seeds of many plants can germinate in dark but seeds of some plants like lettuce and some varieties of tobacco require light for germination. In the absence of light, plants seedlings cannot synthesize chlorophyll and other pigments. They become pale with long internodes, slender and weak stem. This phenomenon is called etiolation. Under prolonged dark conditions seedlings die.
5. **Temperature.** Plants require a suitable temperature for growth. Most of the plants grow in temperature ranging from 4°–35°C. Plants grown at higher altitude are adapted to grow at sub zero temperature conditions and those growing in xerophytic conditions in deserts can tolerate temperature as high as 45°–50°C. However most of the plants grow their best at temperature ranging between 25°–30°C.

Growth Regulators

Growth of plants is controlled by some naturally occurring organic compounds known as phytohormones or plant hormones. Plant hormones are synthesized in minute amount in the growing region of plants. These plants hormones migrate from the region of their synthesis to other plant part to initiate physiological activities leading to promotion, inhibition or some other desired change in plants growth. Plant hormone like auxins, gibberellins and cytokinins are growth promoters whereas abscisic acid is a growth inhibitor. A gaseous plant hormone ethylene brings about ripening of fruits. Different growth hormones may work together in a synergistic manner or in an antagonistic manner.

Auxins

These are a group of hormones and include indole acetic acid (IAA) and indole butyric acid (IBA). IAA was first isolated from human urine and in plants it is synthesized in the shoot apex and root apex. IAA brings about many physiological changes in plants like cell elongation, root initiation and development of fleshy fruits. Auxins also check premature fall of leaf and fruit and causes apical dominance in plants. Auxins along with cytokinins play a role in cell division in plants. 2–4, D (2, 4–dichloro phenoxy acetic acid) a synthetic auxin is employed in agriculture as a weedicide.

Gibberellins

There are many gibberellins, the most important and well studied is GA_3 (gibberellic acid). GA_3 was first isolated from a pathogenic fungus *Gibberella fujukuroi.* This fungus was first noticed in Japan to cause a disease known as bakane disease in rice (foolish seedling growth of rice), in which infected rice plants remain pale and weak and grow extensively with poor yield. Later on many gibberellins were isolated from plants source which are all acidic in nature and named as GA_1, GA_2 and GA_3 etc. Gibberellins are synthesized primarily in the young leaves and roots of the plants. Physiological role of gibberellins in plants is elongation of stem and leaf expansion. Breaking of seed and bud dormancy and promotion of seed germination is another important role of gibberellins. During germination of grains of cereals like barley, gibberellins stimulates formation of α-amylase responsible for hydrolysis of starch.

Cytokinins

These are group of hormones which promotes cytokinesis *i.e.,* cell division along with auxins. Credit of discovering cytokinins goes to **Skoog** and co- workers. They found that kinetin obtained from autoclaved DNA of herring sperms help in cell division. Zeatin is a natural occurring cytokinin found in unripe maize grains. Cytokinins are synthesized in developing shoot buds, root apices, young fruits and endosperm of seeds. Process of cell division, cell elongation and differentiation of cells in plants takes place by the influence of cytokinins and also retards ageing of leaves.

Abscisic Acid

Abscisic acid (ABA) is a growth retardant and occurs in dormant buds. It promotes leaf fall and also causes seed dormancy in some plants. It reverses many physiological activities induced by GA_3 and cytokinins. It also decreases the synthesis of RNA and proteins and promotes ageing of leaves. It also inhibits the formation of α-amylase in germinating grains of cereals.

Ethylene

It is a gaseous hormone and is synthesizes in the tissues undergoing senescence and ripening of fruits. It brings about horizontal growth of seedlings and apical hook formation in dicot seedlings. It promotes senescence and abscission of plant parts like leaves, flowers etc. It is highly effective in inducing fruit ripening.

Exercises

1. What are auototrophs? What is the significance of an auototrophic system?
2. Name and categorize various essential elements required for plant growth.
3. What are the different criteria of essentiality of minerals to the plants?
4. What are deficiency symptoms? How it can be demonstrated?
5. Describe the role of Nitrogen, Phosphorous and Potassium in plant growth.
6. Write a brief note on micronutrients.
7. What is nitrogen fixation? Describe the role of *Rhizobium* in nitrogen fixation in leguminous plants.
8. What is nitrification and denitrification?
9. What is photosynthesis? What are its two phases?
10. Describe the role of chloroplast in photosynthesis. Name the various light absorbing pigments present in it.
11. With the help of graph explain absorption spectrum of chlorophyll – a.
12. Describe the process of light harnessing in plants and explain the process of light reaction of photosynthesis.
13. Describe Calvin cycle.
14. Write in brief about Hatch and Slack pathway.
15. Differentiate between C_3 and C_4 plants.
16. Write a note on environmental factors influencing the process of photosynthesis.
17. Describe the process of plant growth.
18. What are the requirements of plant growth?
19. Write a brief note on plant growth regulators.

❑❑❑

Chapter 5 ANIMAL SYSTEM

Animals are the heterotrophic organisms. There are more than a million types of different animals inhabiting diverse environmental conditions and are present both in water and on land. Animals which occur in water are called aquatic animals and those occurring on land are called terrestrial animals. Aquatic animals occurring in seas and oceans are called marine *e.g.*, star fishes, marine fishes, penguins, prawns, sharks, dolphins, whales. Aquatic animals living in rivers, ponds and lakes are called fresh water forms *e.g.*, fresh water fishes, hippopotamus, crocodile and water birds. Terrestrial animals may live in thick and dense forests like lions, tigers and elephants or may live in underground burrows like rabbits, rats or may live in crevices like snakes and insects or on branches of trees like monkeys, chimpanzee and some can fly in the sky like birds and bats. Many of the animals like cows, buffaloes, goats, sheep, horses and donkeys have been domesticated by man to obtain milk, meat, wool, and for the purpose of joy riding or as drought animals. Cats and dogs are pet animals. Similarly silk worms and honey bees are reared to obtain silk and honey respectively. Most of the animals are either aquatic or terrestrial some like frogs are amphibians *i.e.*, they share both of these habitats.

Body of the animals with the exception of sponges is of constant shape and size with definite symmetry. Different type of symmetry of animals is shown in Fig. 5.1.

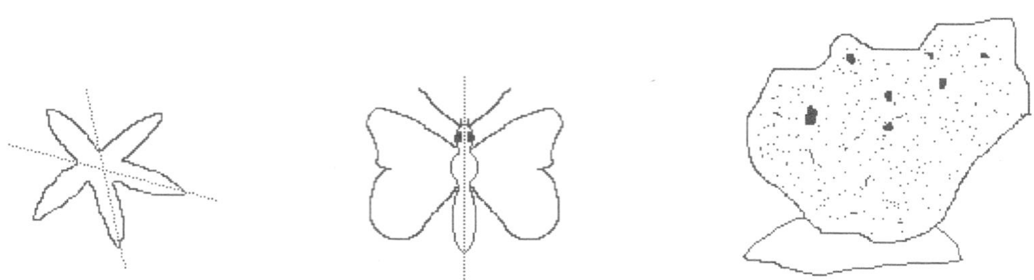

(*a*) (*b*) (*c*)

(*a*) Star fish (radial symmetry), (*b*) Butterfly (bilateral symmetry), (c) Sponge (asymmetric)

Fig. 5.1. Body symmetry of animals.

Majority of them have a bilateral symmetry *i.e.*, their body can be divided into two equal halves by only one plane *e.g.*, butterfly, spiders, ants, cats, dogs. Some animals like star fish and sea anemone have a radial symmetry *i.e.*, their body can be divided into two equal halves by more than one plane. Growth in animals is limited and is proportional to all body parts. It takes place by the enlargement of body organs. Body of higher organisms consists of easily recognizable regions like head, neck, thorax, abdomen and tail. However tail in humans is a rudimentary organ and is represented by a coccygeal region which has four small vertebrae at the hind end of the vertebral column. Animal body also posses paired appendages like forelimbs and hind limbs that are used for locomotion and as a food capturing devices. Appendages may be modified for various type of motion like swimming, climbing or flying as in fishes, lizards and birds respectively. Most of the higher animals also possess endoskeleton consisting of bones and cartilages to support the body and it also provides site for the attachment of muscles. Lower animals like insects have an exoskeleton as their body wall is composed of chitin in which minerals are deposited. Some other lower animals like earthworm are without a skeleton and have a soft body. Animals respond to the stimulus very quickly as they possess well developed sensory organs and nervous system except sponges. Animals have well developed organ and organ system to carry out vital life functions. They have definite anterior, posterior, dorsal, and ventral body regions; body regions of a fish are shown in Fig. 5.2

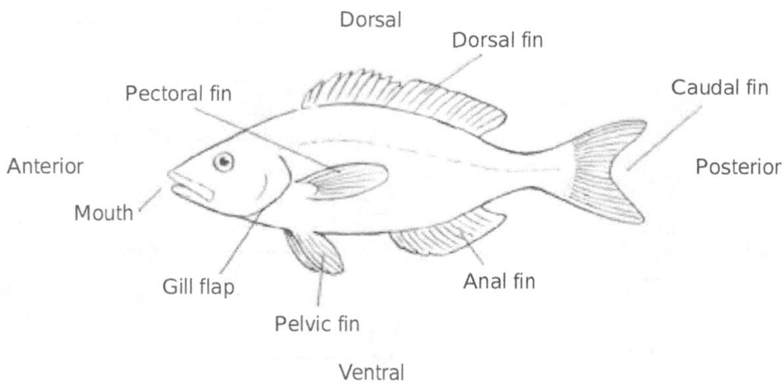

Fig. 5.2. Body regions of a Fish.

In this chapter digestive system, respiratory system, excretory system and circulatory system have been described with special reference to humans.

NUTRITION IN ANIMALS

Food is the basic requirement of the entire living organism. It is required for growth, development, reproduction and maintenance of life. Unicellular animals like amoeba engulf its food with the help of pseudopodia and take it inside its body where food is digested with the help of enzymes. In higher animals mouth is the organ of intake of food (holozoic nutrition). Some parasitic animal like tape worm inhabiting in human intestine do not have a mouth but absorb digested food from the intestine by their body surface. A few other animals like spiders, house fly do not ingest solid food instead they secrete digestive enzymes directly on their food which is usually a dead and decaying organic matter and suck the digested food. Food of some animals is from a plant source and these are called herbivores *e.g.*, cow, camel, horse, goat etc. Some other animals obtain food from an animal source *i.e.* they kill their prey and eat it. These are called carnivores *e.g.*, lion, tiger, python etc. Some other animals are omnivores *i.e.*, obtain their food from both plant and animal source *e.g.*, crow, bear, cockroach. Man is also an omnivore.

HUMAN DIGESTIVE SYSTEM

It consists of alimentary canal and digestive glands. Food is digested in the alimentary canal with the help of digestive juices secreted by the digestive glands. Digestion involves mixing food with digestive juices, moving it through the alimentary canal and breaking down of large molecules of foods into small molecules which are absorbed in the blood and is made available for growth, repair and a source of energy. Human alimentary canal is about 9m long and it is more or less a tubular structure of varying diameter as shown in Fig. 5.3. It starts with mouth and is followed by oral cavity, pharynx, oesophagus, stomach, small intestine, large intestine and ends at anus.

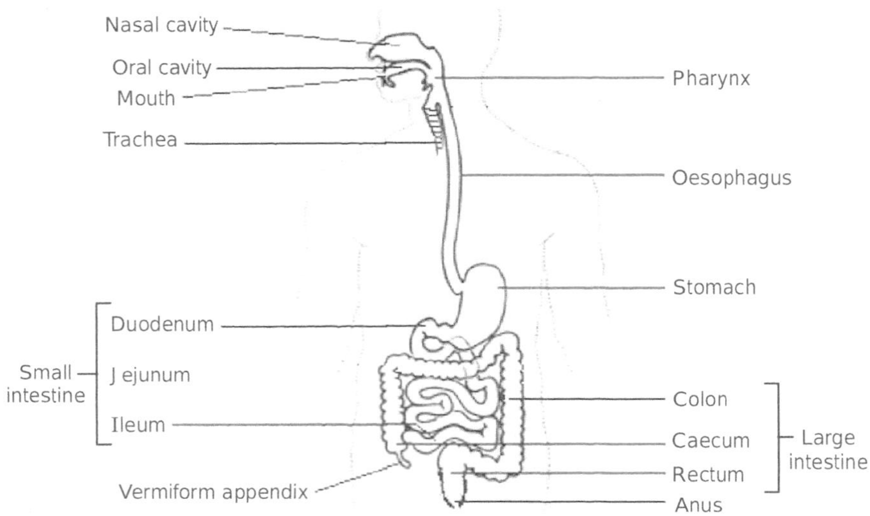

Fig. 5.3. Human alimentary canal.

Digestive glands associated with the digestive system are salivary glands, liver, pancreas, gastric glands and intestinal glands. Specific role of different parts of alimentary canal and the digestive glands in the food digestion is as follows:

1. Mouth

It is the anterior most opening of the alimentary canal present in the head region of the body below the nostrils. It is guarded by upper and lower lips and food is ingested through it. It leads into oral cavity.

2. Oral Cavity

It is a specious cavity and is bounded by upper and lower jaw. Its roof is called palate which is hard in the anterior region and soft towards the posterior side. Floor and side of the oral cavity are supported by throat and cheeks respectively. It has tongue and teeth. Secretion of salivary glands is poured in it. Tongue is highly muscular and slightly protrusible, is attached on the floor of the oral cavity. Human possesses two sets of teeth *i.e.,* milk teeth and permanent teeth. Milk teeth are present during the childhood and are subsequently replaced later on by permanent teeth such teeth are called **diphyodont.** Teeth in adult are present in socket of jaw such teeth are called **thecodont**. Adult human has different type of teeth *i.e.,* incisors, canines, premolars and molars. Such teeth are called **heterodont**. Adult human being has 32 teeth 16 each in upper and lower jaw. In each half of the jaw from middle to the back region are 2 incisors, 1 canine, 2 premolars and 3 molars. Thus the human dental formula is 2123/2123 × 2.

Food digestion is initiated in the oral cavity. It is masticated into small pieces with the help of teeth and is moistened with the help of saliva secreted by salivary glands. There are three pairs of salivary glands one pair in the roof and two pair in the floor of the oral cavity. Tongue taste the food and rotate it to mix saliva in it thoroughly. Saliva contains mucin which lubricates the food and an enzyme called salivary amylase which partially converts starch into maltose sugar. Saliva is active at pH 6.7. Finally food is converted into small balls or bolus which is swallowed into the stomach through pharynx and oesophagus.

3. Pharynx

It is a funnel shaped cavity and is present behind the oral cavity. There is no clear cut demarcation between oral cavity and pharynx. Pharynx leads into oesophagus and through it food passes into oesophagus without any change. Pharynx also opens into trachea (wind pipe) by a separate opening. During swallowing of food epiglottis a cartilaginous flap covers trachea to check entry of food particles in it.

4. Oesophagus

It is a long, narrow, highly muscular and tubular structure. It starts from the pharynx, runs through the neck, thorax and passes through the diaphragm and opens into the stomach, food passes through it without any change.

5. Stomach

It is a J–shaped sack like structure present in the left side of the abdomen and is the widest part of the alimentary canal in which food is stored temporarily. It has thick walls. Food is carried in it in a semi solid manner from the oesophagus. Opening of oesophagus into the stomach is guarded by a cardiac sphincter that checks the regurgitation of food *i.e.,* its back flow into the oesophagus. When food enters into the stomach gastric juice is secreted by its glandular cells. Gastric juice contains hydrochloric acid (HCl), mucus and a protein digesting enzyme trypsin. HCl performs three main functions:

(*i*) It disinfects the food *i.e.,* kills the germs present in the food.

(*ii*) It provides an acidic medium (pH—2) for the activity of trypsin.

(*iii*) It stops the starch digestion which is initiated in the oral cavity as the salivary enzyme is inactivated in an acidic pH.

Mucus protects the inner lining of stomach from the action of HCl and gastric enzymes. For that purpose it is continuously secreted and washes the inner lining of stomach. Trypsin helps in partial digestion of proteins and they are converted into peptones. In the stomach by the muscular action of its wall food is converted into a semi liquid form called chyme. It passes into the duodenum the proximal part of small intestine in small amounts at a time. Opening of stomach into the duodenum is guarded by pyloric sphincter present at their junction and it acts as a valve and allows only the forward flow of food into small intestine.

6. Small Intestine

It is longest part of alimentary canal about 7.5m long and a highly coiled tubular structure present in the abdomen. Internally the wall of intestine is projected into finger like projection called villi that increase the surface area and enhance the capacity of absorption. It consists of three parts: proximal duodenum, middle jejunum and distal ileum.

(*a*) ***Duodenum.*** It is a U shaped tube and receives bile juice from the liver and pancreatic juice from pancreas by a common duct as shown in Fig. 5.4.

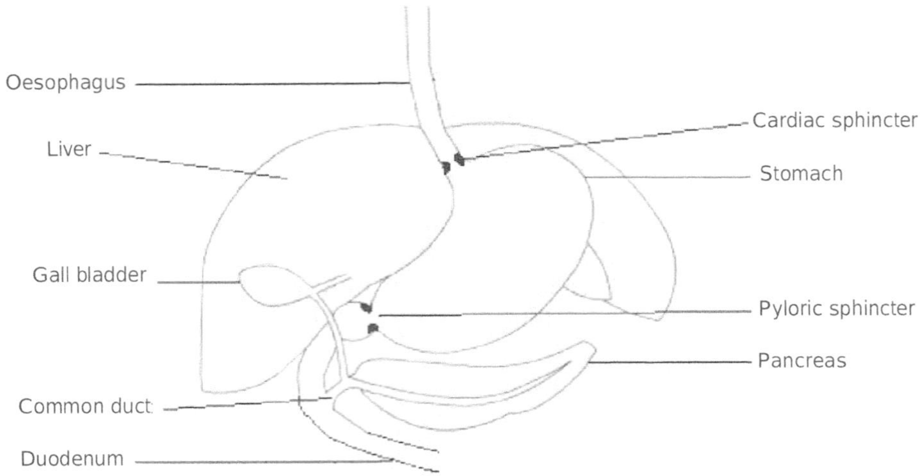

Fig. 5.4. A portion of Human digestive system.

Bile juice is secreted in the liver the largest gland of the human body present in upper part of abdomen. Bile juice is temporarily stored in gall bladder and is poured into the duodenum when it receives food. Bile serves the following function:

(*i*) It provides an alkaline medium for the action of the pancreatic enzyme in the duodenum.

(*ii*) It emulsifies fats *i.e.,* fats are broken into smaller globules.

(*iii*) It brings bile pigments containing remnants of worn out and damaged RBCs for egestion along with the faeces.

Pancreatic juice is secreted by the pancreas present between stomach and duodenum. Pancreatic juice contains three enzyme trypsin, pancreatic amylase and lipase. Trypsin converts proteins and peptones into proteoses, peptides and amino acids. Pancreatic amylase converts starch into maltose. Lipase converts lipids into fatty acids and glycerol.

(*b*) ***Jejunum.*** Partially digested food enters into jejunum the middle part of small intestine where it met with sucus entericus *i.e.,* the intestinal juice secreted by glandular cells of intestine. It contains a number of enzymes like peptidase, maltase, lactase, sucrase etc. Peptidase converts polypeptides and peptones into amino acids. Maltase converts maltose into glucose. Lactase converts lactose the milk sugar into glucose and galactose. Sucrase converts sucrose into glucose and fructose.

(*c*) ***Ileum.*** It is the last part of the small intestine and when food enters into it is almost in a digested form what ever a little undigested food is left is digested in it with the help of same type of enzymes as that of jejunum.

7. Large Intestine

Small intestine is followed by large intestine which is about 1 m long. It is sacculated in appearance and is broader than the small intestine. No digestion of food takes place in it. It consists of three parts *i.e.*, caecum, colon and rectum. Caecum connects ileum the last part of the small intestine with colon and it is extended into a blind small tube the vermiform appendix which is a vestigial organ in human. In caecum cellulose digestion takes place in herbivore ruminants like rabbit, horse etc. not in humans. Colon is the longest part of large intestine and it consists of an ascending limb, a horizontal limb and a descending limb. In colon undigested food is converted into semisolid faeces by absorbing excess of water from it. Faecal matter is passed into rectum where it is stored temporarily and is egested through anus.

8. Anus

It lies at the end of abdomen between two legs. It is an opening present at the hind end of the alimentary canal and is guarded by an anal sphincter and faecal matter is egested through it.

Digested food is absorbed by the blood capillaries present in the wall of the intestine. It is distributed to various body parts through blood vessels. Then it is assimilated in the body *i.e.*, utilized as an energy source and in making of building blocks for new cells.

Respiration in Animals

It is a vital life activity in animals similar to all other living organisms. During respiration digested food, mainly glucose is oxidized in body cells to liberate energy that enables the living organisms to carry out various physiological and physical activities essential for their survival and maintenance. Respiration is of two types: aerobic respiration and anaerobic respiration.

(*i*) ***Aerobic Respiration.*** It takes place in the presence of oxygen and occurs in majority of organisms. During this process complete break down of glucose molecule takes place and it is converted into carbon dioxide and water liberating energy as shown in the equation below.

$$\underset{\text{Glucose}}{C_6H_{12}O_6} + \underset{\text{Oxygen}}{6O_2} \longrightarrow \underset{\text{Carbon dioxide}}{6CO_2} + \underset{\text{Water}}{6H_2O} + \text{Energy (674 kcal)}$$

(*ii*) ***Anaerobic Respiration.*** It takes place in the absence of oxygen and occurs in a few organisms including bacteria, yeast etc. Anaerobic respiration in microbes is known as fermentation. In this type of respiration incomplete break down of glucose takes place as shown in the equation below.

$$\underset{\text{Glucose}}{C_6H_{12}O_6} \longrightarrow \underset{\text{Alcohol}}{2C_2H_5OH} + \underset{\text{Carbon dioxide}}{2CO_2} + \text{Energy (54 kcal)}$$

Anaerobic respiration also occurs in our muscles during vigorous physical activity. At the time of vigorous muscular activity oxygen gets used up faster in our muscles than it is made available to them. In such conditions lactic acid is synthesized in our body by the partial break down of glucose. Accumulation of lactic acid in muscles cause pain and this pain gets subsidized when one rests as during that time sufficient oxygen is made available to the muscles to oxidize the accumulated lactic acid.

The energy liberated from glucose in aerobic respiration and anaerobic respiration is 674 kcal/mol and 54 kcal/mol respectively. Thus the energy liberated in aerobic respiration is much more as compared to anaerobic respiration. In majority of animals aerobic respiration occurs and in this process animals obtain O_2 from the environment and liberate CO_2 from their body to the outside environment. This exchange of respiratory gases is the function of their respiratory system and is carried out in two phases: external respiration and internal respiration.

(*i*) ***External Respiration.*** During the external respiration O_2 from outside environment diffuses into respiratory organs and in turn CO_2 is expelled out from the later to outside environment.

(*ii*) ***Internal Respiration.*** During the internal respiration O_2 from the surrounding extracellular fluid diffuses into the body tissues for oxidation of food and CO_2 is liberated during the process which is eliminated from tissues to the surrounding extracellular fluid. Transfer of O_2 from respiratory organs to tissues and elimination of CO_2 from tissues to the respiratory organs takes place through the blood.

There are different types of respiratory organs like skin, trachea, gills and lungs found in different type of animals. In most of the aquatic animals like fishes and prawns gills are the respiratory organs. Gills are bathed continuously with water as it is drawn onto the gills by the movement of fish's mouth coupled with those of gill covers. Oxygen present in water diffuses in

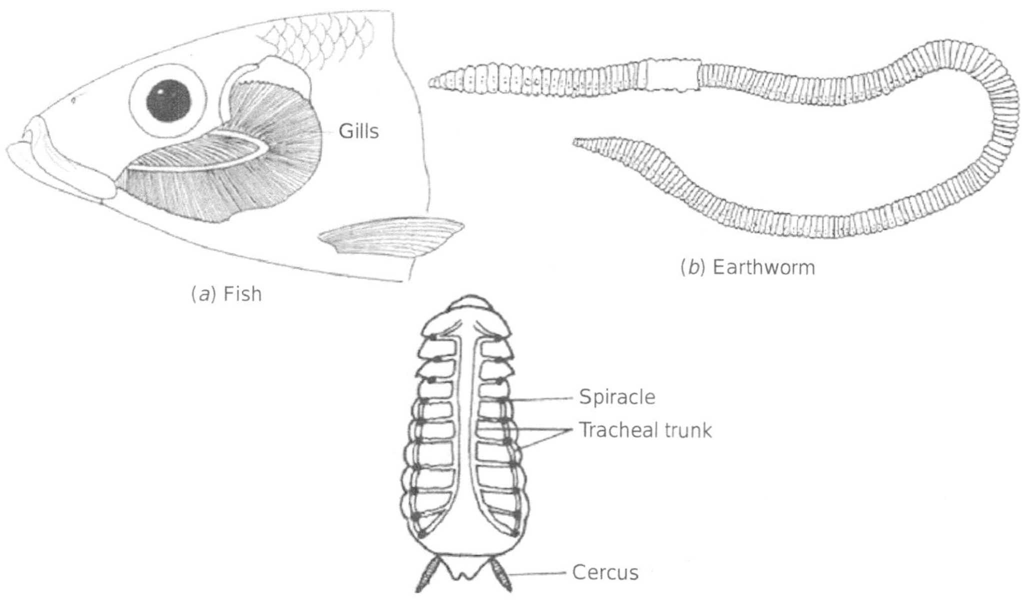

Fig. 5.5. Respiratory organs.

blood capillaries of gills and carbon dioxide is liberated out from them into the outside water. Majority of terrestrial animals like lizards, birds and mammals have lungs as the respiratory organs. Human beings respire through lungs. Respiration through the lungs is called pulmonary respiration. Soft bodied animals like earthworm respire through the skin. Animals that respire through skin have a thin and highly vascularised skin which is always kept moist to facilitate gaseous exchange. This type of respiration is called cutaneous respiration. Frog is an amphibian animal and respires both through skin and lungs. Insects like grasshopper and cockroach respire through the trachea (respiratory tubes).In tracheal respiration exchange of O_2 and CO_2 between body tissue and outside environment takes place directly without the involvement of blood. Tracheal system is such that trachea reach every part of the body. Oxygen rich air enters through the openings called spiracles situated on the sides of the body of an insect and reaches to all tissue directly through trachea and their branches *i.e.,* tracheoles. Conversely carbon dioxide is diffused out from tissue to outside environment through the spiracles.

Above mentioned respiratory organs like lungs, gills, skin and trachea have large surface area. They are thin walled to get easily and enough oxygen from outside environment and to diffuse out carbon dioxide from the body.

HUMAN RESPIRATORY SYSTEM

Human respiratory system (Fig. 5.6) consists of a respiratory tract and a pair of lungs (respiratory organs). Primary function of the respiratory system is to obtain oxygen for the use of body cells from the outside environment and elimination of carbon dioxide that produce into the environment.

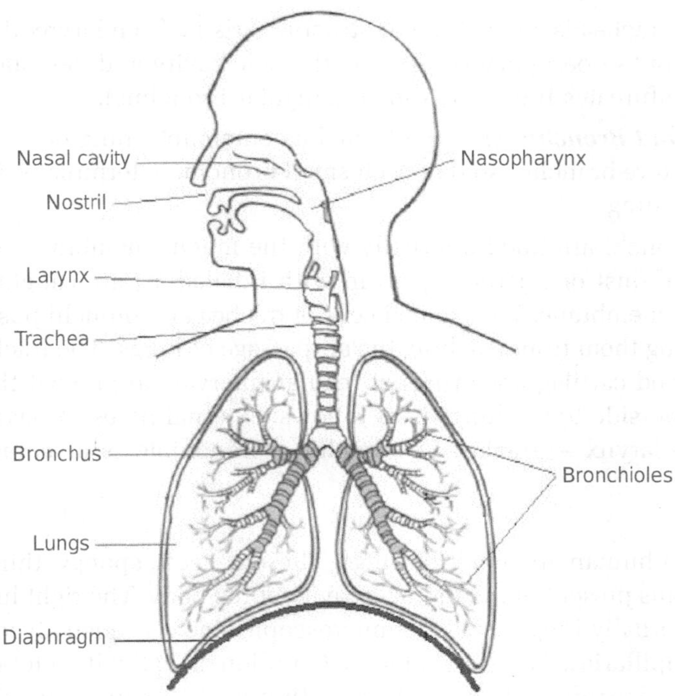

Fig. 5.6. Human respiratory system.

Respiratory Tract

It consists of external nares, nasal cavity, internal nares, pharynx, larynx, trachea, bronchi and bronchioles. Function of the respiratory tract is intake of fresh air containing oxygen from environment to the lungs and elimination of foul air containing carbon dioxide from lungs to the outside environment.

(*i*) ***External Nares* (nostrils).** These are the two semicircular openings; present at the tip of nose, each of which enter into a nasal cavity.

(*ii*) ***Nasal Cavity.*** Incoming air is moistened, warmed and filtered in the nasal cavity. Inside it are present number of hair and its internal lining has mucus secreted by nasal glands in which germs and other suspended air particles are entrapped when air moves towards the inner side. Nasal cavity leads into pharynx via internal nares.

(*iii*) ***Internal Nares.*** These are two narrow slits in the posterior part of wall of pharynx which communicates nasal cavity with pharynx.

(*iv*) ***Pharynx.*** Internal nares open into posterior part of pharynx known as nasopharynx. It is a common passage both for food and air.

(*v*) ***Larynx (voice box).*** Larynx is a tubular chamber supported by cartilaginous framework. Opening of pharynx into larynx is guarded by a cartilaginous leaf like flap called epiglottis which covers this opening during food swallowing. Besides its role in respiration, it is also a sound producing organ. It has two vocal cords that are folds of its membrane and they vibrate when air strike those producing sound. Larynx opens into trachea.

(*vi*) ***Trachea.*** Trachea is a long tubular structure arising from larynx that runs through the neck in front of oesophagus, into the thoracic cavity and descend down up to lungs where it bifurcates into two bronchi (singular bronchus).

(*vii*) ***Bronchi and Bronchioles.*** Each bronchus enters into lung of its respective side and branch and re branches so that each small bronchiole terminates in an alveolus or air sac of the lung.

Trachea and bronchi are lined internally with the mucus membrane which is kept moist. Any small particle of dust or microbes pass in with inhaled air are caught and held in moist surface of the mucus membrane. The ciliated cells of trachea and bronchi push up these particles and help in preventing them from reaching the air passage of lungs. The trachea and bronchi are supported by C shaped cartilaginous rings at regular intervals to prevent their collapsing. The passage of air from outside to the lungs is as follows: external nares → nasal cavity → internal nares → pharynx → larynx → trachea → bronchi → bronchiole → alveolus of the lungs.

Lungs

Respiratory organs in human are a pair of lungs. These are soft, spongy, thin walled and highly vascular pinkish organs present in the thoracic cavity of the body. The right lung is slightly bigger than the left one. Internally lungs consist of microscopic air sacs called alveoli which are served by network of fine capillaries. There are about 300 million alveoli with a total surface area equal to $100m^2$.Each lung is surrounded by two layered thin walled membrane called pleura. A fluid is present in between the pleura that cushions the lungs and prevents friction. Lungs are well protected inside the thoracic cavity by ribs from the sides, ventrally by sternum and dorsally by vertebral column. Towards the posterior side they are protected by the diaphragm; a muscular partition that divides the body cavity into anterior thoracic region and a posterior abdominal region.

MECHANISM OF RESPIRATION

Respiratory system performs the function of exchange of respiratory gases between environment and human body. During this process intake oxygen present in air for its use to liberate energy by oxidizing food substrate in body tissues and out take of carbon dioxide formed during food oxidation is eliminated out of body into the environment. Flow chart of respiratory gases exchange between environment and human body is shown in Fig. 5.7. This process is divisible into four steps: breathing or ventilation, pulmonary gas exchange, gas transport through blood and gas exchange in tissues.

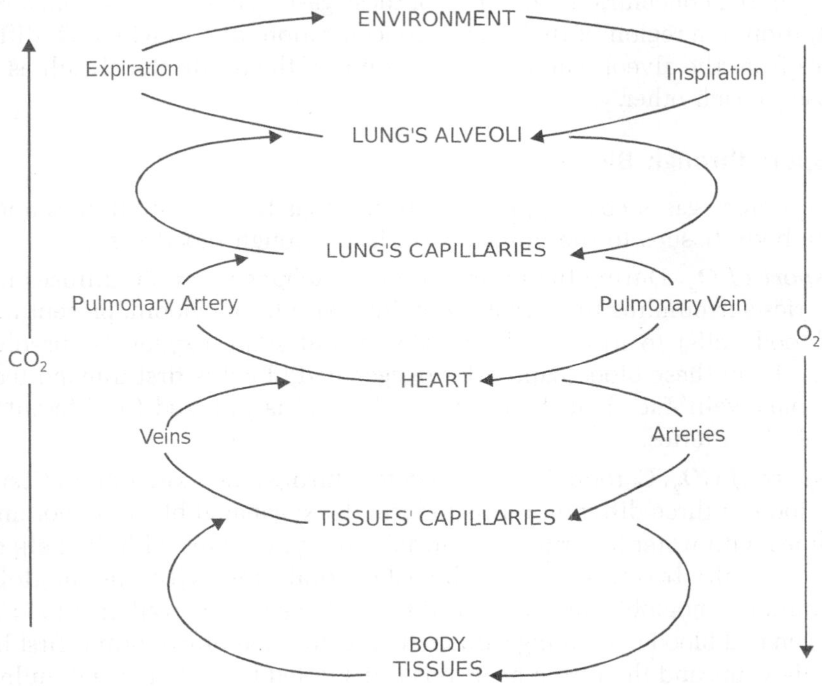

Fig. 5.7. Flow chart of respiratory gas exchange.

(*i*) Breathing or Ventilation

It includes inspiration *i.e.*, breathing in of fresh air loaded with oxygen from outside environment to the lung's alveoli and expiration *i.e.*, breathing out of foul air loaded with carbon dioxide from the alveoli to the outside environment.

(*a*) ***Inspiration.*** It occurs when the thoracic cavity expands. Thoracic cavity expands when the diaphragm is lowered and becomes flat from its arched upraised position by contraction of its muscles coupled with moving of ribs upward and outward by the contraction of inter costal muscles. It results in decrease of air pressure inside the thoracic cavity as a result of which fresh air containing O_2 rushes into the lung's alveoli through nostrils, trachea and bronchi.

(*b*) ***Expiration.*** When the muscles attached to the ribs and diaphragm relaxes the ribs are lowered and move inward; diaphragm is raised and becomes arched as a result of which thoracic cavity compressed forcing out the foul air containing CO_2 out of lungs via trachea through the nose.

Each inspiration and expiration constitutes a single breathe. A normal human being breathes 12 to 14 times per minute in a resting position and the rate increases to 18 to 20 times per minute during vigorous exercise. Though breathing is an involuntary exercise its rate is regulated by the respiratory centre of the brain.

(*ii*) Pulmonary Gas Exchange

Lung's alveoli are surrounded by blood capillaries and deoxygenated blood is supplied to them by pulmonary artery from the heart. As lung's alveoli are filled with O_2 rich air obtained by inspiration and their capillaries contain CO_2 in the deoxygenated blood; exchange of gases occurs due to difference in the concentration gradient of these gases. Gases moves from a region of their higher concentration to a region of their lower concentration; due to which O_2 diffuses into the blood capillaries from the alveoli and CO_2 diffuses out of them into the alveoli as they are thin walled and close to each other.

(*iii*) Gas Transport through Blood

Transport of respiratory gases *i.e.*, supply of O_2 from lungs to the body tissues and elimination of CO_2 from the body tissues to the lungs takes place through blood vessels.

(*a*) ***Transport of O_2.*** During the pulmonary gas exchange; oxygen diffuses into the blood capillaries surrounding the alveoli; combines with haemoglobin, present in erythrocytes (red blood cells) to form oxyhaemoglobin and some oxygen is dissolved in blood plasma. From these blood capillaries oxygenated blood is first transported to heart via pulmonary vein and then the oxygenated blood is pumped to different body tissues through the arteries.

(*b*) ***Transport of CO_2.*** Carbon dioxide liberated during food oxidation in tissues combines with blood in three different forms to form deoxygenated blood. Major amount of CO_2 combines with water to form carbonic acid in erythrocytes with the help of an enzyme carbonic anhydrase, some of the CO_2 combines with haemoglobin to form carbaminohaemoglobin and a very little of it gets dissolved in blood plasma. The deoxygenated blood containing carbon dioxide in a combined form is first brought to the heart via veins and then from heart it is transported to lungs through pulmonary artery for the elimination of CO_2 during expiration.

(*iv*) Gas Exchange in Tissue

Tissues get oxygen from the blood capillaries supplied by arteries carrying oxygenated blood and oxygenation of blood occurs regularly during inspiration. In tissues O_2 is utilized in cellular respiration to liberate energy in food oxidation and during the process CO_2 is generated as a byproduct which diffuses into the capillaries. This exchange of gases occurs due to difference in the concentration gradient of the respiratory gases. In this manner tissues get regular supply of oxygen from the oxygenated blood and carbon dioxide produced by them is diffused out from the deoxygenated blood through the lungs during expiration.

CIRCULATION OF BODY FLUID IN ANIMALS

Body fluid is circulated in animals for the purpose of transport of nutrients, waste products, respiratory gases, hormones etc. from one body part to the other. To transport substances from one body part to the other almost all the animals possess some mechanisms to circulate the body fluid. However, in some primitive, aquatic and sedentary animals like sponges transport of the substances occur by the circulation of water from their surroundings. They possess an extensive canal system in the body. *Scypha* is a marine sponge. There are numerous pores called ostia on its body surface; water from the surrounding medium enters through these pores, circulates in the canals and leaves the body through a wide opening called osculum. Incoming water carries food and oxygen to the body cells from the surroundings and the outgoing water takes away waste products and carbon dioxide from the body cells to the outside environment. Similarly another aquatic and sedentary animal *Hydra* that possesses central water filled body cavity or coelenterons which communicates to the surroundings by a common opening the mouth or manubrium, through which both inward and outward movement of water occurs. Food, waste products and respiratory gases are exchanged between water and body cells.

Most advance animals posses a well developed circulatory system; consisting of heart and blood vessels to conduct the blood. There are two type of circulatory systems; open and closed. Many invertebrates like insects possess an open circulatory system. In an open system the blood (haemocoel) pumped by the heart comes out from the blood vessels and flows through open spaces and channels in tissues before returning to heart. Some lower animals like leeches, and earthworms etc. and all the vertebrates possess a close circulatory system. In a close circulatory system the blood pumped by the heart remains confined to the blood vessels. It never leaves the blood vessels unless injured. Humans possess a close circulatory system.

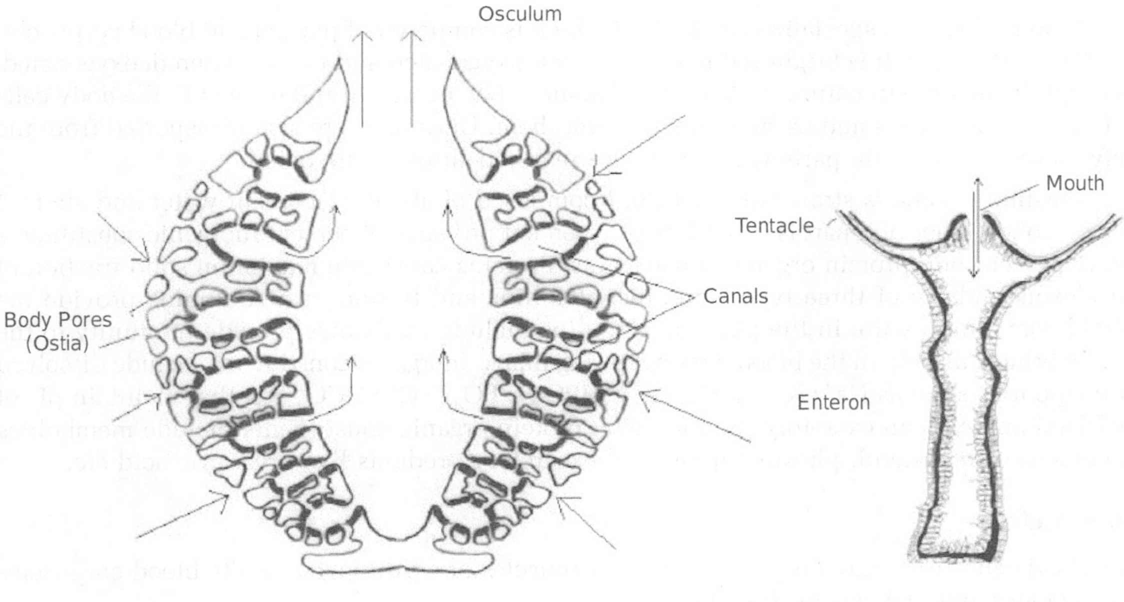

(*a*) Sponge (Scypha) (*b*) Hydra

Fig. 5.8. Coarse of water circulation in primitive aquatic animals

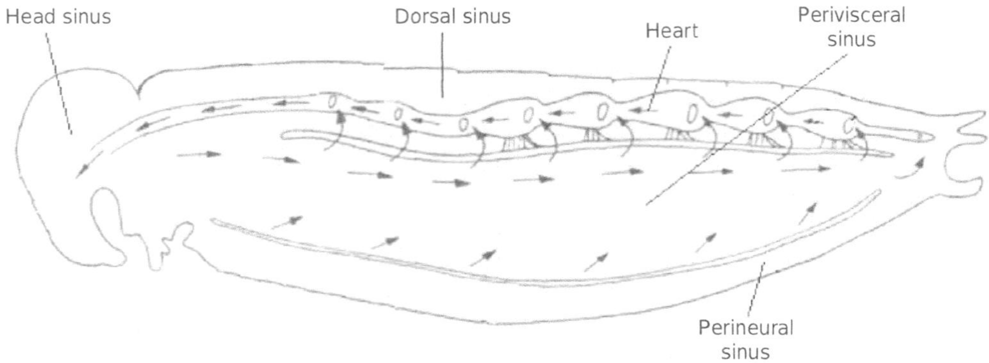

Fig. 5.9. Circulatory system of Cockroach (open type).

HUMAN CIRCULATORY SYSTEM

It consists of two interconnected systems: blood vascular system and lymphatic system.

Blood Vascular System. It comprises of blood, heart, and blood vessels. Its main function is transport of food, respiratory gases, hormones and waste products from one organ to another in the body.

Blood. Blood is a specialized body fluid which is composed of plasma and blood corpuscles or cells floating in it. It is bright red in colour when oxygenated and purple when deoxygenated. It is slightly alkaline in nature. It delivers substances like nutrients and oxygen to the body cells and takes away wastes and carbon dioxide from them. Hormones are also transported from the body glands to the body parts where they are required through the blood.

Plasma. Plasma is straw coloured fluid composed of about 92 percent water and about 8 percent solids. The solid part is consisting of 7 percent proteins, .9 percent inorganic constituents and .1 percent non protein organic constituents. Proteins constitute maximum solid fraction of the plasma and are of three type albumins, globulins and fibrinogens. Albumins provide the colloid osmotic pressure in the plasma. Globulins include antibodies provide immunity to the body. Fibrinogens help in the blood clotting during injury. Inorganic constituents include dissolved inorganic ions such as Na^+; K^+; Ca^{2+}; Mg^{2+}; HPO_4^{2-}; PO_4^{3-}; Cl^-; HCO_3^- etc. that maintain pH of the blood and NH_3 an excretory product. Non protein organic constituents include metabolites like glucose, cholesterol, phospholipids; and execretory products like urea, uric acid etc.

Blood Cells

The blood cells are of three types: red blood corpuscles or erythrocytes, white blood corpuscles or leucocytes and platelets or thrombocytes.

(a) Red Blood Cells

The red blood cells (RBC) or erythrocytes are circular, biconcave and disc shaped cells possessing no nucleus in the mature stage. They are produced in bone marrow and have a life span of 120 days. They impart red colour to the blood. They possess an iron containing pigment called haemoglobin which transports oxygen from lungs to body tissues and bring back carbon dioxide from the tissues to the lungs. In a normal healthy individual there are about 4.5 to 5.5 million

RBC per cubic millimeter of blood. When RBC counts decreases it results in anaemia in which haemoglobin concentration is below normal.

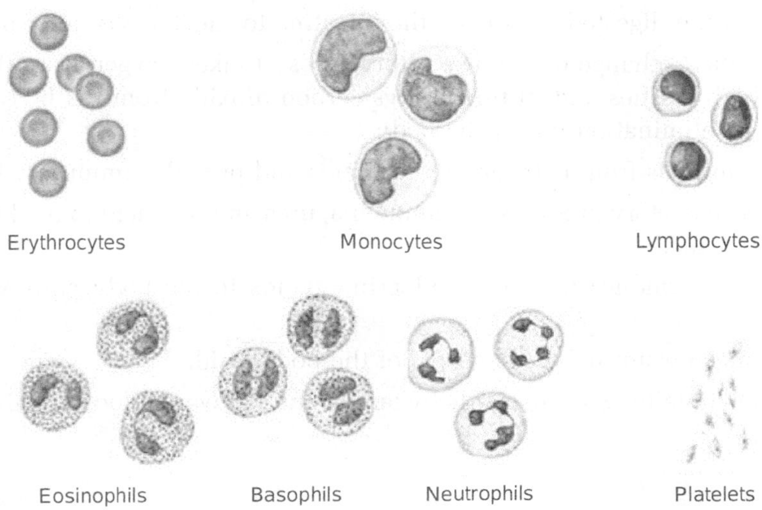

Fig. 5.10. Different type of Human blood cells.

(b) White Blood Cells

The white blood cells (WBC) or leucocytes are colourless and are rounded or irregular in shape. They change their shape and show amoeboid movement. But unlike red blood cells they possess a nucleus. They are fewer in number but larger than RBC. There are about 6000 to 10000 WBC per cubic millimeter of blood in a normal healthy individual. They are produced in bone marrow and in lymph glands. Their main function is to destroy harmful bacteria and to remove some other foreign materials from the tissues and blood by ingesting them *i.e.,* eating them. They are of two main types: agranulocytes and granulocytes.

(*i*) ***Agranulocytes.*** They are without granules in the cytoplasm and possess unlobed nucleus and are produced in the red bone marrow. These are of two sub types: lymphocytes and monocytes. The monocytes have kidney shaped nucleus whereas lymphocytes possess a rounded nucleus. Lymphocytes produce serum globulin and play an important role in immunological reactions. Monocytes eat bacteria and other foreign matter by phagocytosis.

(*ii*) ***Granulocytes.*** They possess granules in their cytoplasm and their nuclei are bilobed. They are of three sub types: eosinophils, basophils and neutrophils. The eosinophils have coarse granules in their cytoplasm and are stained with acidic dyes such as eosin. They play a role in detoxification. The basophils have fewer coarse granules in the cytoplasm which are stained with basic dyes such as methylene blue. They have two or three lobed nucleus. They play a role in allergic reactions. The neutrophils do not take any colour when exposed to acidic or basic dyes. They protect us against infection.

(c) Blood Platelet

They are very small irregular bodies smaller than RBC and do not have a nucleus but their cytoplasm contains distinct granules. Their number ranges from 2,50,000 to 4,00,000 per cubic millimeter of the blood. They play an important role in blood clotting.

Functions of Blood

The important functions of blood in animals are:

1. It transports the digested food from the intestine to the various body parts.
2. It helps in the exchange of the respiratory gases. It takes oxygen from the lungs to the different body tissues and in turn brings carbon dioxide from the body tissues to the lungs for its elimination out of the body.
3. It protects animals from pathogenic organisms and provides immunity to the body.
4. It transports excretory products like ammonia, urea and uric acid to the kidneys for their removal as urine.
5. It transports hormones from the endocrine glands to the body parts where they are required.
6. It regulates the water balance and pH of the body fluid.
7. It helps in maintaining a constant body temperature of warm blooded animals (birds and mammals).

Blood Clotting

When an injury ruptures a blood vessel, blood flows out of it. After sometime, blood flow stops as it thickened to form a blood clot. This is a natural preventing mechanism of the body which prevents excessive loss of blood from an injured body part. Blood clotting occurs as follows:

As blood flows out of a damaged blood vessel, blood platelets release a substance called thromboplastin. In the presence of thromboplastin and calcium ions, prothrombin present in the blood plasma is converted into thrombin. Thrombin then catalyzes the conversion of soluble plasma protein fibrinogen into an insoluble one fibrin. The fibrin molecules quickly polymerizes into long fibres which form a mesh in which blood cells get entangled. A crust thus formed is called a clot which plugs the injury and stops bleeding. The clot later on shrinks and squeezes out a yellowish liquid called serum. It is plasma minus fibrinogen.

Heart

Structure of human heart is shown in Fig. 5.11. It is a reddish coloured muscular organ of conical shape present between the two lungs in the thoracic cavity. Its broad end is called the base which is present towards the upper side and its pointed end is termed as apex lies towards the lower side facing the diaphragm. It is situated almost in the middle of chest but its apex is tilted to the left. It is enclosed in a two layered transparent sac called pericardium. The space between two layers of the pericardium is filled with a pericardial fluid that keeps the heart moist and facilitates free and frictionless movement. The heart beats spontaneously and rhythmically throughout the life. The heart of a normal human beats 72 times per minute. It has four chambers: two atria or auricles situated towards the upper broader end and the two ventricles present towards the lower conical end. The right and left halves of the heart are completely separated by a partition called septa. The right and left auricles are separated by an inter-auricular septum. An oval depression is present on this septum known as fossa ovalis that represents remnants of foramen ovale; an opening in inter auricular septum of the foetal heart. The two ventricles are separated by inter ventricular septum.

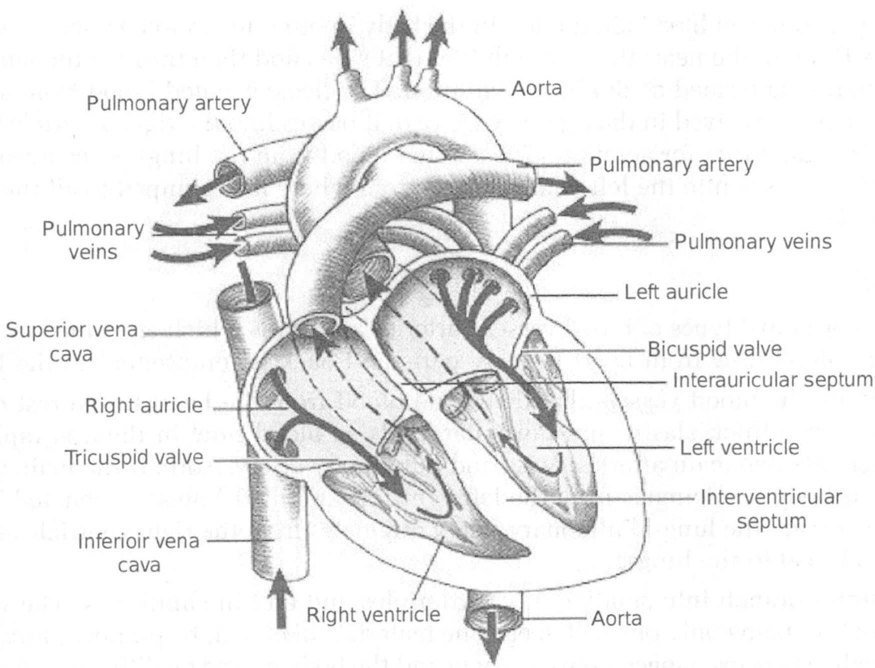

Fig. 5.11. Internal structure of Human heart.

Each auricle opens into the ventricle of its own side through an auriculo-ventricular aperture. The two apertures are guarded by valves that allow flow of blood from auricles into ventricles and prevent its back flow. Opening of left auricle into left ventricle is guarded by Mitral valve or bicuspid valve having two flaps. A tricuspid valve consisting of three flaps guards the opening of right auricle into right ventricle. Like wise, at the base of the major arteries that arise from the ventricles and distribute blood to the body parts have semilunar valves which prevent the back flow of blood into the ventricles.

Pumping Action of Heart

The heart pumps the blood to the various parts of the body. The walls and its chambers are made up of cardiac muscles and they contract and relax in a rhythmic manner. This rhythmic contraction followed by its relaxation constitutes a heart beat. The contraction of a chamber decreases its volume and forces the blood out of it whereas its relaxation increases its volume to receive more blood. Contraction of heart's chamber follows a regular order; first the right auricle contracts then the left auricle, there after the right ventricle and the left ventricle contracts in the last. The rhythmic beating of heart maintains circulation of blood and transport substances from one part of the body to the other part. There is a complete separation of oxygenated and deoxygenated blood in the two halves of the human heart. The right auricle receives deoxygenated blood from all the body parts except the lungs through two main vessels; a superior vena cava and an inferior vena cava; bringing blood from the upper regions and the lower regions respectively. The left auricle receives oxygenated blood from the lungs through four pulmonary veins. The ventricles are highly muscular particularly the left ventricle and receive blood from the auricle of their respective sides through auriculo ventricular apertures. The right ventricle pushes the deoxygenated blood into large pulmonary trunk which soon divides into two arteries leading to the respective lungs. The left ventricle pumps the oxygenated blood into thick aorta or the main distributing artery that distributes the blood to the rest of the body parts. When the ventricles relax and regain original size they draw more blood from the auricles of their respective sides.

During the course of blood circulation in the body in order to circulate once round the body it flows twice through the heart first through the right side , and then through the left side . This type of circulation is termed as **double circulation.** The deoxygenated blood from all the body parts except lungs is received in the right auricle then it passes into the right ventricle from where it is pushed into the lungs for aeration. The aerated blood from the lungs is received by the left auricle and then passes into the left ventricle and from where it is pumped to all the body parts except the lungs.

Blood Vessels

There are two principal types of blood vessels: arteries and veins which are joined by capillaries. Circulation of blood flow from heart to body parts and back is represented in the Fig. 5.12.

Arteries are the blood vessels that distribute blood from the heart to the rest of the body parts. Arteries have thick elastic and muscular walls as blood flow in them is rapid and at a pressure. There are two main arteries: aorta and pulmonary artery. Aorta is the main distributing artery of the as body it originates from the left ventricle and distributes oxygenated blood to all the body parts except the lungs. Pulmonary artery originates from the right ventricle and supplies deoxygenated blood to the lungs.

The arteries branch into small arteries, arterioles and end in capillaries. The walls of the capillaries are thin being only one cell deep. The materials like food, respiratory gases, hormones and waste products are exchanged between them and the body organs by diffusion. The capillaries lead into small venules which unite to form small veins and finally small veins unite to form large veins.

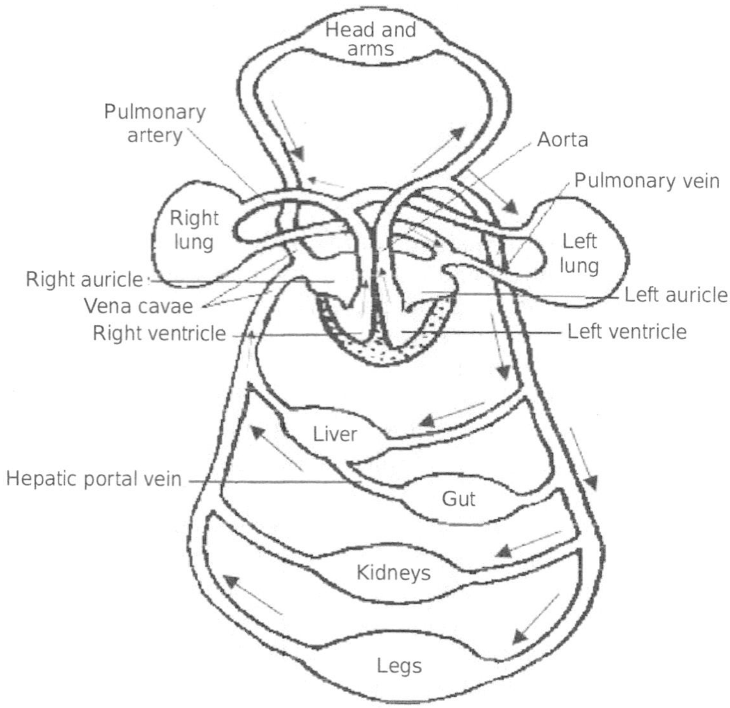

Fig. 5.12. Diagramatic representation of blood circulation in Human.

Veins are the blood vessels that carry blood from the various body parts to the heart. Veins are thin walled, blood flow in them at low pressure, and have valves at regular intervals to prevent back flow of blood. There are two major veins superior vena cava and inferior vena cava which brings deoxygenated blood from all the body parts except lungs to the right auricle of the heart. Pulmonary veins bring oxygenated blood from the lungs to the left auricle.

Lymphatic System

The lymphatic system is consists of a colourless fluid, the lymph, lymphatic capillaries and lymphatic vessels, lymph glands and lymph nodes.

Lymph is a light yellow fluid containing lymphocyte cells. Lymph is like blood but it lacks red blood cells and a few blood proteins. Lymph is present around the body tissue in the intercellular spaces. It acts as a carrier and helps in shuttling the materials between body tissues and the blood. It takes food, oxygen and hormones from the blood and gives them to the tissues and in turn takes waste materials and carbon dioxide from the tissues and gives it to the blood. Lymph flows only in one direction that is from the tissues to the heart.

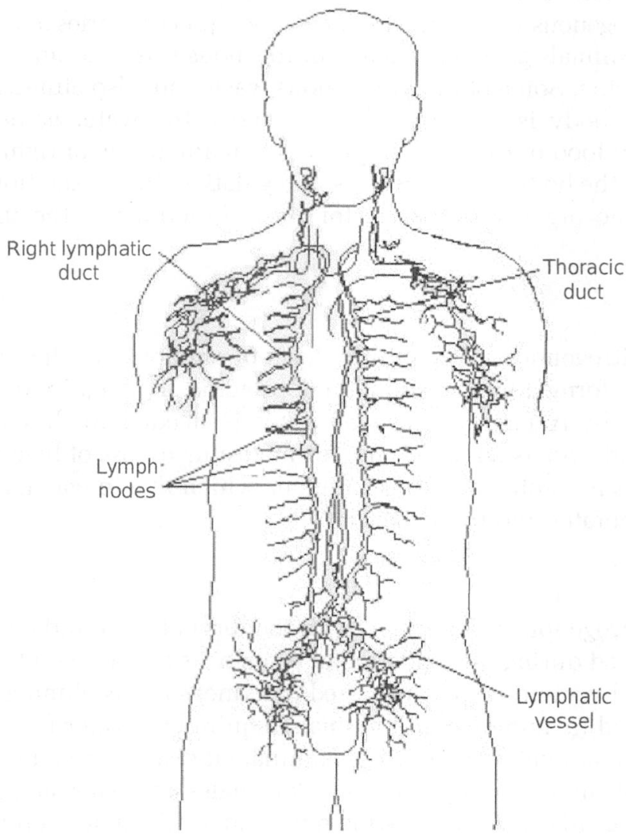

Fig. 5.13. Human lymphatic system.

From the intercellular spaces lymph enters into the lymph capillaries which are interwoven with blood capillaries around the body tissues. Lymphatic capillaries unite to form lymphatic vessels. Lymph vessels are similar to veins except that there walls are thin and have more valves.

On their way to heart lymph vessels they come across with a number of lymph nodes at regular intervals. Lymph nodes are minute sac like structures consisting of lymphoid tissue in which the foreign bodies (germs) are filtered and destroyed by the lymphocytes. The lymphatic vessels unite to form large vessels which in turn unite to form two main lymphatic vessels or trunks called left thoracic duct and right lymphatic duct.

The left thoracic duct receives the lymph from the entire body except right arm, and right side of head, neck and thorax. The left thoracic duct opens into the vein near the junction of left internal jugular vein and the subclavian vein. The right lymphatic duct receives lymph from right arm right side of the head and remaining part of thorax which opens in the vein at the junction of right subclavian and internal jugular vein. The lymphatic system provides an accessory route for the flow of inter cellular fluid into the blood.

Excretion in Animals

Excretion means elimination of waste materials particularly the nitrogenous waste from the body formed during biochemical reactions. These are mainly produced by catabolism of proteins. Elimination of wastes from the body is must otherwise their accumulation in the body would be toxic. The nature of nitrogenous wastes and their excretion process varies among animals depending on the water viability. Animals generally excrete nitrogenous wastes as ammonia, urea or uric acid chiefly in the form of urine. Some of the nitrogenous wastes are also eliminated through the body surface. Moreover, the body is also required to remove the water gained in excess of body requirement either from food or drinks. The process of maintaining of right amount of water and proper ionic balance in the body is termed as osmoregulation. Both excretion and osmoregulation go side by side and the organ responsible for these functions is the urinary system or the excretory system.

Ammonotelism

It means removal of nitrogenous wastes in the form of ammonia in the urine. Ammonia is the basic nitrogenous waste formed by the oxidation of proteins. It is highly soluble in water and also highly toxic to the body. A large amount of water is needed to dissolve ammonia and its elimination from the body. So its elimination in urine results in loss of high amount of water from the body. But this poses no problem to those animals which live in water. Ammonotelic animals include aquatic invertebrates and bony fishes.

Ureotelism

It means removal of nitrogenous wastes mainly in the form of urea in the urine. In such animals ammonia which is formed during the oxidation of proteins is converted into urea by reacting with carbon dioxide in their liver and then transported to kidneys for its elimination along urine. Urea is less toxic as compared to ammonia and animals require less water for its removal. Further, it can be concentrated and animals can afford to eliminate it at a slower rate. Examples of ureotelic animals are all mammals including aquatic ones like whales and seals, amphibians like toads and frogs, cartilaginous fishes like sharks and sting rays, and semi aquatic reptiles like alligators and turtles. Man is also ureotelic and can concentrate urea in urine more than hundred times as compared to its concentration as present in blood.

Uricotelism

It means elimination of nitrogenous wastes chiefly in the form of uric acid. In the uricotelic animals, ammonia formed by the breakdown of proteins is converted first into uric acid in their liver. Uric acid is then transported to kidneys for its elimination outside the body along the urine. Uric acid is less toxic as compared to ammonia and urea. Conversion of ammonia into uric acid requires expenditure of energy but that is compensated in these animals which include birds, reptiles and insects as they have limited access to water and uric acid is insoluble in water and thus its elimination requires a little of water.

HUMAN EXCRETORY SYSTEM

Human excretory system (Fig. 5.14) is consists of a pair of kidneys, a pair of ureters, a urinary bladder, and a urethra.

(*i*) Kidneys

Kidneys are the main excretory organs in human. Kidneys separate the wastes from the blood. There is a pair of kidneys located in the abdominal cavity one on each side of vertebral column just below the diaphragm. The left kidney is usually placed a little above the right one. Each of the kidneys is somewhat bean shaped with concavity towards the inner side.

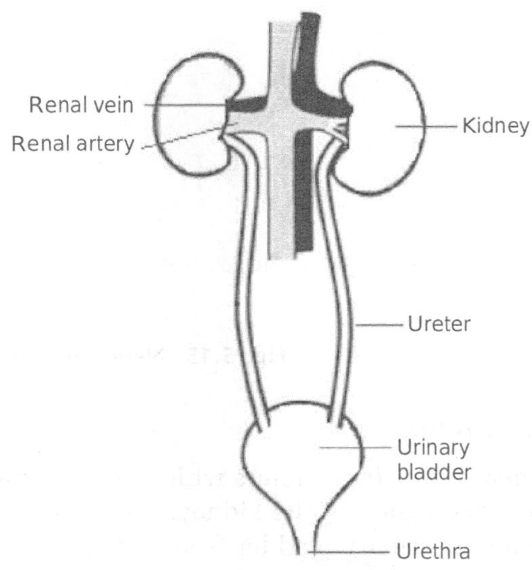

Fig. 5.14. Human excretory system.

Each kidney receives blood from a renal artery. This blood carries nitrogenous and other wastes. Each kidney contains several tiny excretory units called nephrones in which filtration of blood and urine formation takes place. The blood free of wastes leaves from each kidney by a renal vein.

Structure of Nephron and Filtration of Blood

Nephron (Fig. 5.15) is a structural and functional unit of kidney. Each of the nephron is composed of a cup shaped structure called Bowman's capsule which leads into a tubular part consisting of a proximal convoluted tube, U-shaped Henle's loop and a distal convoluted tube. It is surrounded by a number of blood capillaries. Filtration of blood and urine formation takes place in it.

The waste materials are brought to kidney by the renal artery. Each of the nephron receives a narrow afferent arteriole arising from the renal artery.Afferent arteriole splits into a bundle of capillaries called glomerulus in the Bowman's capsule where blood is filtered. The filtered blood free of wastes is passed on to efferent arteriole,and it breaks into capillaries and the later unite to form a renal veinule. Renal veinules of all nephrones unite to form renal vein which takes the filtered blood out of kidney to the heart.

The filtrate carrying the wastes passes through the tubular part where useful products such as glucose, amino acids, some salts and considerable amount of water are reabsorbed by blood capillaries surrounding the nephrones and wastes from capillaries are poured in it by counter current mechanism; details of which are beyond the scope of this book. Waste from the nephrones of each kidney is passed into collecting ducts. Collecting ducts unite to form tubules which open into ureter of their side.

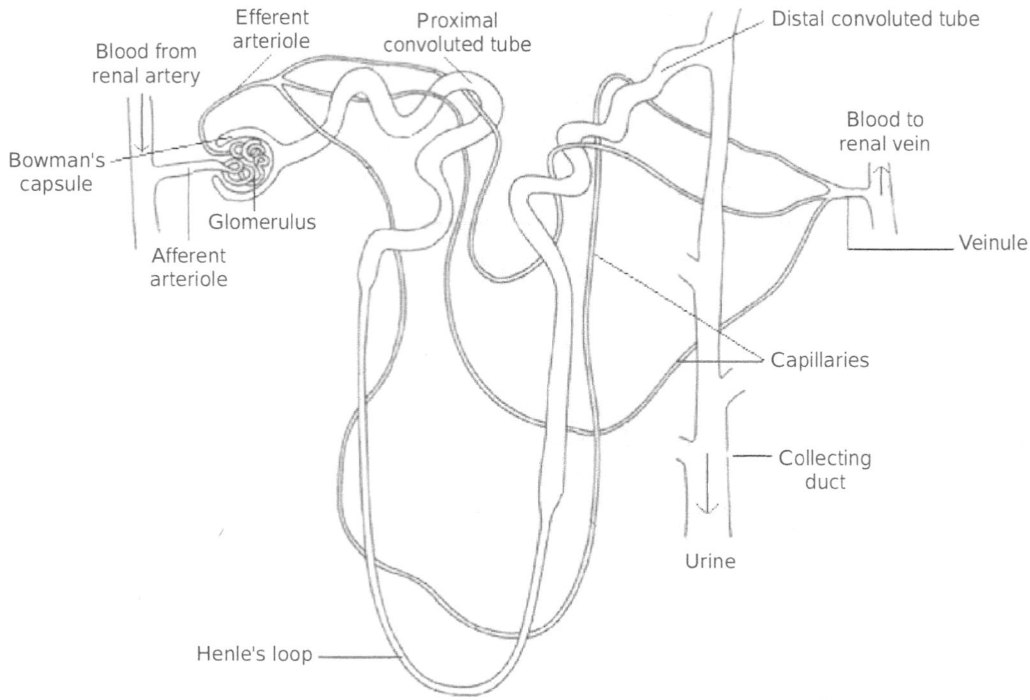

Fig. 5.15. Nephron-Structural and functional unit of kidney.

(*ii*) Ureters

There is a pair of ureters which are thin muscular tubes each of which arise from middle of the concave region of the kidney; they descend down and open into urinary bladder. Ureters carry urine to urinary bladder from kidneys.

(*iii*) Urinary Bladder

Urinary bladder is transparent membranous sac which is highly muscular and distensible organ that lies on the pelvic floor in the lower abdomen. It collects urine excreted in the kidneys. Urine is stored in the bladder temporarily. Urine is a transparent, yellowish, aqueous liquid. It is acidic in nature and is more concentrated as compared to blood plasma. It has a characteristic smell. Urine contains water and nitrogenous wastes most of which is urea, uric acid, ammonia, creatinine and some other wastes. When bladder is filled with urine it is expelled out of body through urethra. The act of voiding the urine is called micturition.

(*iv*) Urethra

Urethra is a narrow tube which connects the urinary bladder and urine is expelled out of body through it. Urethra performs the excretory function in both male and female; passes out urine to outside and also reproductive function in male; as a passage for semen. Urethra in females opens between clitoris and vaginal opening whereas urethra in males opens at the tip of penis.

Osmoregulation by Kidneys

Osmoregulation means regulation of water content and ion concentration in the body. Kidneys not only function in the elimination of toxic waste, but also control water balance of body. Excess of water in the body increases blood volume and its pressure. As a result of which filteration of blood in glomerulus is at a pressure and the filtrate has more water. Excess of water in the body also inhibits the release of pituitary hormone vasopressin which results in less reabsorption of water in capillaries surroundings the nephrones consequently more water is eliminated in the urine. If the amount of body fluid falls below normal, glomerulus filteration slows down and there is more reabsorption of water in the capillaries due to the release of vasopressin and as a result of which urine becomes more concentrated.

Accessory Excretory Organs

In addition to kidneys which are the major excretory organs in human the other organs which assist in excretion are: the skin, lungs, and liver. These organs also eliminate metabolic waste products directly or indirectly.

(*i*) ***Role of Skin in Excretion.*** Human skin possesses sweat glands that secrete sweat. Sweat is an aqueous fluid mainly composed of water containing some sodium chloride and traces of urea and lactic acid and helps in the elimination out of body.

(*ii*) ***Role of Lungs in Excretion.*** Lungs eliminate carbon dioxide formed in the body by the oxidation of food substrates chiefly the carbohydrates.

(*iii*) ***Role of Liver in Excretion.*** Liver changes the haemoglobin of worn out red blood corpuscles into bile pigments (bilirubin and biliverdin). These pigments are passed into the intestine and are eliminated along with faeces. Conversion of ammonia into urea also occurs in liver and is transported to kidney for elimination.

Exercises

1. Describe the human digestive system with the help of a well labeled and suitable diagram.
2. Write in brief about nutrition in animals.
3. Explain in brief about aerobic and anaerobic respiration in animals.
4. Write a note on different type of respiratory organs in animals.
5. Describe the human respiratory system with the help of a well labeled and suitable diagram.
6. Write a note on open and closed type of circulatory system in animals. What are the functions of circulatory system?
7. Describe the composition and functions of human blood.
8. What is the significance of blood clotting?
9. Describe the structure of human heart.
10. Write a note on human lymphatic system.
11. What is the importance of excretory system?
12. Write a note on ammonotelism, ureotelism and uricotelism.
13. Describe the human excretory system.
14. Describe the structure of nephron. How blood filtration and urine formation occurs in it?

Chapter 6 MICROBIAL SYSTEM

An organism which can not be seen by an unaided eye is called a *microorganism*. They can be observed with the help of a microscope and the study of microorganisms is called *microbiology*. Microorganisms are found almost everywhere on the earth. The microorganisms or microbes do not belong to any single group of organisms according to the available systems of classification but are an assemblage of diverse type of organisms that differ from each other in many respects and the only point of similarity in them is that they are too small and require a microscope to observe them. The major types of microorganisms are: archaea, bacteria, mycoplasma, cyanobacteria, protists, fungi, and viruses.

HISTORY OF MICROBIOLOGY

It was **Antony van Leeuwenhoek** (Fig. 6.1), a Dutch merchant, who in the year 1676, observed microorganisms using a single lens microscope designed by him and named them *'animalcules'*. Later on they were found to different microbes like bacteria, fungi, algae and protozoa. Although Leeuwenhoek (1632–1723) had a little formal education but glass cutting and grinding was his hobby and this lead him to design a simple microscope with the help of which he had a glimpse of microbial world first of all. With the discovery of bacteria and other microorganisms by Leeuwenhoek a new branch of science—microbiology came into existence. He is considered as *'Father of Microbiology'*. However the name bacterium was coined much later by **Ehrenberg** in 1828. The field of bacteriology is generally considered to have been founded by **Feridinand Cohn** (1828–1898), a botanist whose studies on algae and photosynthetic bacteria led him to describe several bacteria including *Bacillus* and *Beggiatoa*. Cohn was also the first to formulate a scheme for the taxonomic classification of bacteria.

With the discovery of bacteria a debate started regarding their origin. At that time there was a widespread belief of spontaneous creation of life. According to this view, life arises from the life less decaying organic matter or spoiled food such as beef broth. **Lazzaro Spallanzani** (1729–1799) disproved the spontaneous creation of life with the help of experiments with broth in 1768. Spallanzani demonstrated that microorganisms found in decaying food stuff is due to their presence in the air, which settle on the broth and thus grows their. He took two sets of vessels containing broth. One of it was left open in the air, and the other was properly sealed after boiling the broth in it to kill any bacteria that might already be present. Spallanzani observed that no bacteria appeared in the sealed vessel and appeared only in the vessel exposed in the air. But the proponents of spontaneous creation remained unconvinced, they argued that boiling the broth destroyed some *'vital principle'* present in the air and due to this region no microbe appeared on the sterilized broth. The theory of spontaneous creation was finally refuted in a convincing manner by **Louis Pasteur**.

Fig. 6.1. Antony van Leeuwenhoek (1632–1723).

Fig. 6.2. Louis Pasteur (1822–1895)

Fig. 6.3. Robert Koch (1843–1910).

Louis Pasteur (Fig. 6.2) helped in establishing the science of microbiology and immunology and expounded the germ theory of diseases. He ended the long held belief that microorganisms arise spontaneously from non living substances during the process of decay (putrefaction) with the help of his famous *'swan neck flask'* experiment in 1862 as shown in Fig. 6.4. Pasteur supported Spallanzani's view point and also redress the criticism leveled at his experiment, namely boiling might destroy *'vital principle'* in air. He devised a long *'swan necked flask'*.

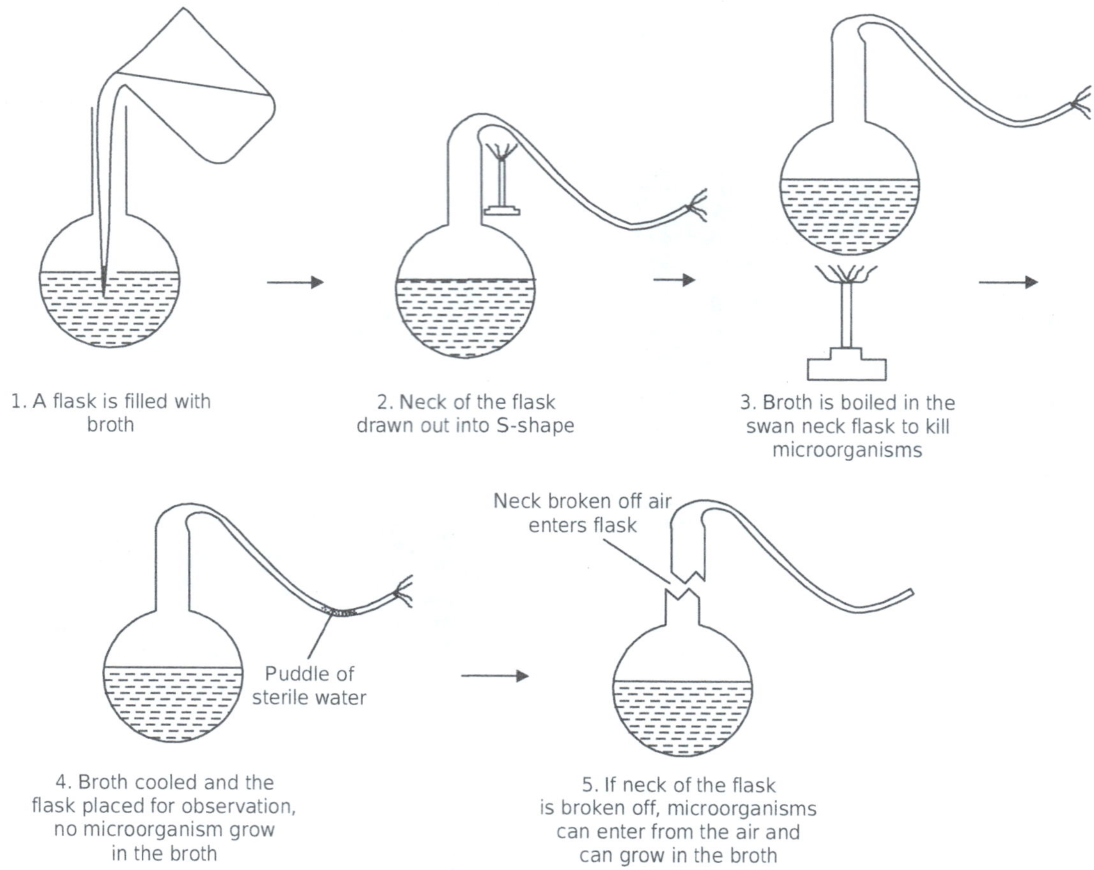

Fig. 6.4. Demonstration of Pasteur's swan neck experiment.

Air could reach the flask through the opening but dust particles and microorganisms could not, because the curved neck served as a trap. He boiled broth in a swan necked flask to ensure that no microorganism survived in the broth as well as in the neck and after boiling it was cooled and left as such. He observed that the broth remains sterile for long time in the flask demonstrating that there is no *'vital principle'* in the air. He explained that no microbes grew in the broth because bacteria present in the air could not climb through the bent tube. He also explained that on breaking the neck of the flask, microbes present in the air settle on the broth and grow there.

He carried out extensive work on fermentation and is also known for the discovery of vaccines against several diseases such as anthrax, fowl cholera and rabies. He also formulated methods of food preservation (pasteurization). Pasteur along with Robert Koch are considered to be founders of medical microbiology.

Robert Koch (Fig. 6.3) is best known for his contribution to the germ theory of diseases proving that specific diseases were caused by specific pathogenic microorganisms. He gave Koch's postulates; with the help of which pathogenicity of a pathogen *i.e.*, the disease causing agent (a bacterium, a fungus, a protozoan, or a virus) can be established. Koch's postulates are:

(*i*) The pathogenic microorganisms must be found in abundance in all organisms suffering from the disease.

(*ii*) The microorganisms must be isolated from the diseased organism and grown in pure culture.

(*iii*) The cultured microorganisms must reproduce the disease when injected into a healthy organism.

(*iv*) The microorganism must be reisolated from the inoculated experimentally diseased organism and identifies as being identical to original specific disease causing agent.

These postulates remain useful even today and are applied by the pathologists to establish pathogenicity of a disease causing agent. Koch's postulates have contributed significantly to the development of basic principles of microbiology. Koch was one of the first scientists to focus on the isolation of bacteria in pure culture. He also found that the cause of tuberculosis is a bacterium, *Mycobacterium tuberculosis*.

Establishment of germ theory of diseases by Pasteur and Koch lead many researchers to work in this field. Prominent among them were **Emile Roux** and **Emil Behring** who discovered diphtheria antitoxin about the year 1894. **Theobald Smith** discovered in 1893 germs of Texas fever in blood corpuscles of cow. **Joseph Lister** (1827–1912), a surgeon from England developed the technique of antiseptic surgery. In 1929 **Alexander Fleming** discovered the antibiotic penicillin which proved to be an effective remedy for many dreadful infectious diseases.

Martinus Beijerinck (1851–1931) helped in establishing the field of virology; he explained the nature of causative agent of tobacco mosaic disease which was first noticed by **Adolf Meyer** a German botanist in 1886 who reported that this disease could be transmitted to healthy tobacco plants by applying the leaf extract of a diseased plant. **Iwanowski** (1882), a Russian scientist observed the same result by applying sap of diseased leaves after passing through bacteria proof filter paper. Beijerinck (1896) considered these filterable agents to be fluid in nature and called them as *Contagium Vivum Fluidum.* Later on **W.M.Stanely** (1935) of the USA obtained the causative agent of mosaic disease of tobacco in a crystalline form and is known as TMV (Tobacco Mosaic Virus). Beijerinck also developed enrichment culture technique which helped in cultivation of wide range of microbes. Beijerinck along with **Sergei Winogradsky** (1856–1953), are considered as founders of general microbiology as they worked on different aspects of microbiology.

Winogradsky was the first to develop the concept of chemolithotrophy and revealed the essential role played by microorganisms in geochemical processes. He first of all isolated and described both nitrifying and nitrogen fixing bacteria.

TYPES OF MICROBES AND THEIR PROPERTIES

There are various types of microorganisms like archaea, bacteria, mycoplasma, cyanobacteria, protists, fungi, and viruses.

1. Archaea

Archaea are a group of single celled microorganisms and are bacteria look alike which are treated as *'living fossils'* that are providing clues to the earliest forms of life on the earth. In the past they were viewed as unusual group of bacteria and named archaebacteria, but since the archaea have independent evolutionary history and show many differences in their biochemistry from other life forms they are now treated as a separate group of prokaryotic organisms. Archaea are similar to bacteria as they lack a well defined nucleus and all other membrane bound cell organelles are absent in them. But their cell wall is unique and consists of proteins and polysaccharides. Cell membrane of archaea contains branched chain lipids and this enables them to tolerate extremes of heat and pH. They reproduce by binary fission, fragmentation and budding but unlike bacteria and eukaryotes they do not form spores.

Archaea includes methanogens, halophiles , thermophiles and psychrophiles. Methanogens are strict anaerobes and produce methane from carbon dioxide or formic acid. They are found in marshy areas; and in the cattle's rumen and thus helping them in the digestion of cellulose. The methanogens are responsible for production of methane gas in biogas fermenters. Thermophiles are aerobes and occur in hot sulphur springs tolerating a temperature around 80°C. These organisms oxidize sulphur to sulphuric acid. Such an environment becomes highly acidic (pH-2). The sulphur dependent archaea are termed as thermoacidophiles. Some archaea occur in extremely strong brine (salt solution). They are called halophiles (salt loving). Psychrophiles are the archaea occurring at sub zero temperature in ice caps.

2. Bacteria

Bacteria are the prokaryote unicellular organisms having different shapes and size. Generally a bacterial cell has an average size of .1μm to .3 μm. Common forms of bacteria (Fig. 6.5) are spherical (*Coccus*, pl. *Cocci*), rod shaped (*Bacillus*, pl. *Bacilli*), comma like (*vibrio*), or spiral (*Spirillum*, pl. *Spirilli*). Their cell wall consists of peptidoglycan. They lack a well defined nucleus and most are without chlorophyll. Many are motile as they possess flagella with the help of which they swim when present in fluid medium. Bacteria multiply by fission and a bacterial cell can replicate into two in just 30 minutes under favourable environmental conditions. In adverse conditions many bacteria can remain dormant forming highly resistant spores with thick protective coverings. Bacteria may be aerobic or anaerobic. They are found almost everywhere where existence of life is possible. They are present in soil; water as well as the air. One gram of garden soil may contain as many as 10^9 bacteria and constitute approximately half of the microbial biomass.

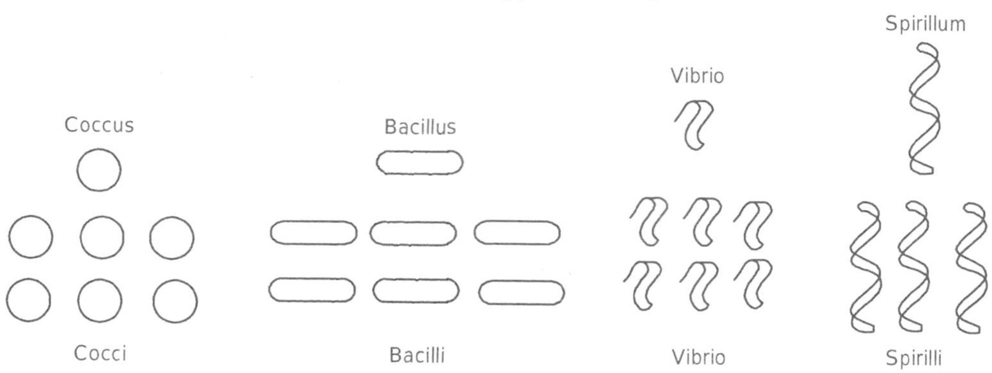

Fig. 6.5. Forms of bacteria.

Majority of the bacteria are saprophytes deriving their nourishment growing on dead and decaying organic matter. Saprophytic bacteria are decomposers and act as natural scavengers; converting the complex organic matter into simple inorganic constituents; and thus help in mineral recycling. Some of the bacteria are parasitic and derive their food growing on other living organisms, such bacteria are pathogenic causing diseases in plants, animals including human beings. Some bacteria occurs in symbiotic association with other organisms *e.g., Rhizobium* occur in the nodulated roots of legumes deriving food from them and in turn provides ammonia and amino acids to the legumes. Similarly bacterium *Escherichia coli* occur in human intestine as a symbiont. Some bacteria contain a form of chlorophyll and carry out photosynthesis and are thus autotrophic. Autotrphic bacteria do not use water (H_2O) as a source of protons for the reduction of carbon dioxide into carbohydrates instead use other sources of protons such as hydrogen sulphide (H_2S).Photosynthetic bacteria are of two types: photoautotrophs and chemoautotrophs obtaining energy from light and chemical reactions respectively for the reduction of carbon dioxide (CO_2).

Actinomycetes are the filamentous bacteria present in dry and warm soil. They are particularly abundant in the soil rich in decomposed organic material. Some well known actinomycetes are *Streptomyces, Micromonospora and Nocardia*. They are decomposers and some of them produces antibiotics which check the growth of pathogenic organisms.

3. Mycoplasma

Mycoplasmas are the smallest unicellular prokaryotic organisms. They are also called as pleuropneumonia like organisms or **PPLO**. Their cell is without a cell wall and thus their body can change shape easily. They are parasitic organisms and cause several diseases in domestic animals (pleuropneumonia) and plants (witches' broom).

4. Cyanobacteria

Cyanobacteria are also known as blue green algae **(BGA)** and are the autotrophic photosynthetic organisms occurring in aquatic habitat. *Nostoc, Spirulina* and *Oscillatoria* are important members of cyanobacteria. These are the prokaryotic organisms and a cyanobacterial cell is normally larger than the bacterial cell. Some of the cyanobacteria are unicellular and others are colonial forming few or many celled filaments. Some cells of cyanobacteria are differentiated into a special type of cell called heterocyst which performs the function of nitrogen fixation. Nitrogen fixing cyanobacteria enrich the nitrogen content of the environment where they grow. In paddy field nitrogen fixing blue green algae plays a significant role in enriching the nitrogen contents of the soil. Nitrogen fixing cyanobacteria like *Nostoc* form symbiotic relationship with a number of primitive and advanced plants like *Anthoceros* a bryophyte, *Azolla* an aquatic fern and *Cycas* a gymnosperm. Cyanobacteria are the earliest oxygenic photosynthetic organisms and helped in making the earth's environment oxidizing from the reducing one.

5. Protist

Protists include all unicellular eukaryotic microorganisms chiefly inhibiting in aquatic habitat. The term protist was coined by **Ernst Haeckel** in 1866. A protist cell body contains a well defined nucleus containing the genetic matter DNA organized in chromosomes and have membrane bound cell organelles like mitochondria, endoplasmic reticulum, Golgi body etc. Protists are a diverse group of microorganisms which includes: unicellular algae, protozoa, and slime moulds.

(*i*) ***Unicellular Algae.*** There are several types of unicellular algae like dinoflagellates and Diatoms. They are collectively called *phytoplanktons* or microscopic floating photosynthetic organisms. They possess a cellulosic cell wall which is heavily impregnated with silica. These are photosynthetic organisms and are found in aquatic habitat. Dinoflagellates have flagella for swimming whereas diatoms are free floating. They are the chief producers in oceans and fresh water environment and accounts for 80 percent of carbon dioxide fixed in the biosphere.

(*ii*) ***Protozoa.*** Protozoa are one celled animals and are the smallest of all animals. Their cell membrane is referred as pellicle. They are found in diverse type of habitats and are cosmopolitan in distribution. Protozoa play a key role in maintaining the balance of bacteria, algae and other microbial life.

Some of them are holozoic in nutrition, and are predatory which feed other protists ingesting particulate food like. These include *Amoeba* and *Paramecium* occurring in freshwater; and *Olpidium* occurring in marine water. *Amoeba* does not have a definite body shape as it frequently changes its shape by producing pseudopodia, the folds of body extension, and with the help of which it captures its prey and also moves around. Paramecium has cilia for locomotion and it ingests food through peristome (oral groove). *Olpidium* is a predatory marine form that inhibits all seas.

Others have a parasitic mode of nutrition *e.g., Plasmodium* and *Entamoeba* are parasitic in human causing malaria and dysentery respectively.

Interestingly, *Euglena* a flagellated protozoan shows myxotrophic mode of nutrition. It can eat food like an animal do (partly heterotrophic) and can make food like plants do (partly autotrophic) as it possesses chlorophyll.

(*iii*) ***Slime Moulds.*** Slime moulds are peculiar type of microorganisms occurring in damp places. They exhibit characters of both fungi and animals. Their body consists of slimy mass of protoplasm with many nuclei of various colours. It is called plasmodium which lacks a cell wall. It slowly streams or glides over decaying leave and engulfs food particles; bacteria and other microbes. In the feeding stage it is like an animal. Under dry condition when food supply is exhausted the plasmodium changes, taking the appearance of a fungus fructification. During such condition the plasmodium transforms into a spore bearing 'fruiting body' like a fungus and this transformation occurs within hours. Examples of slime moulds are *Physarum* and *Stemonites.*

6. Fungi

Fungi are eukaryotic unicellular as well as multicellular heterotrophic organisms and their cell wall is usually composed of chitin. Fungal cell possess a well defined nucleus and other membrane bound cell organelles as are present as in a plant cell except chloroplasts. Body of a fungus may consists of a single cell as in yeast(*Saccharomyces*) or fine thread like structures called *hyphae.* Hyphae of some fungi like *Mucor* and *Rhizopus* are multinucleate without any septa. Such hyphae are called coenocytic hyphae. Hyphae of fungi like *Aspergillus* and *Fusarium* are septate. Hyphae are collectively known as mycelium. Fungi reproduce vegetatively by fragmentation of mycelium, asexually by formation of spores or conidia, and sexually by formation of perfect spores. However, some fungi never reproduce sexually; these are called fungi imperfecti *e.g., Colletotrichum* and *Cercospora.*

Majority of the fungi are saprophytes and act as decomposers. They grow on decaying organic matter in moist and warm conditions and help in mineral recycling. Some of the fungi are parasitic. *Puccinia* and *Ustilago* are parasitic fungi on wheat causing rust and smut disease respectively. Well known unicellular fungus yeast which brings about fermentation of sugars converting it into alcohol and carbon dioxide. It is the basis of brewing and baking industries. Another important fungus is *Penicillium* which is a source of antibiotic penicillin.

7. Viruses

Viruses (Fig. 6.6) are the sub microscopic acellular entities consists of a protein coat and either of the nucleic acids **DNA** or **RNA**. These are obligate parasites which can replicate only inside the cells of their host. These are a unique group of microorganisms which behave as living organisms when present in their host and can be stored as any chemical substance once outside of their host. Crystals of a virus can be stored in a bottle for several years. Viruses infect all types of living organisms: bacteria, plants, and animals including human beings. Viruses which infect bacteria are known as *bacteriophages*. Viruses are about 100 times smaller than a bacterium and they can be viewed with the help of an electron microscope.

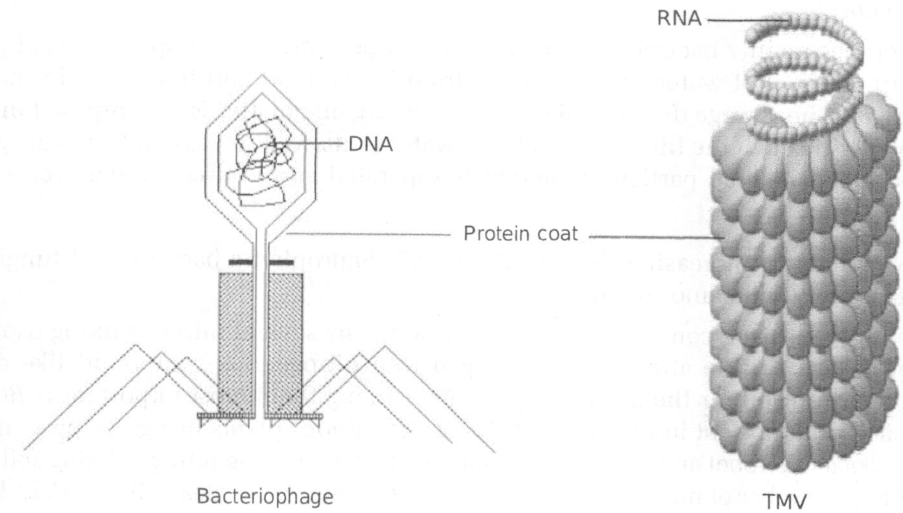

Fig. 6.6. Viruses.

Viruses do not have any machinery of their own but utilize the host cell's machinery (enzymes, nucleotides, amino acids, and energy rich molecules like ATP) when they replicate in their host. Viruses can not be cultured on a synthetic medium without a living cell. Common examples of viruses are: T2 phage infecting bacterium *E.coli*, TMV infecting tobacco plant, and Influenza virus infecting human.

ECONOMIC IMPORTANCE OF MICROORGANISMS

Microorganisms are of great economic importance. They exhibit both beneficial and harmful activities.

Beneficial Activities

There are many beneficial activities of microorganisms. They are an important component of the ecosystem. Most of the microbes like saprophytic bacteria and fungi are decomposers and play a significant role in the maintenance of ecological balance. Some photosynthetic microbes like algae are the chief producers of the aquatic ecosystems and their presence is must for the viability of these systems. Microbes like bacteriophages are an important tool employed in biotechnology. The other useful activities of microbes are:

1. Decay and Decomposition

Decay and decomposition is one of the most important activities of microorganisms chiefly the saprophytic bacteria and fungi. They help in disposing off the litter and act as natural scavengers. Microbes convert the complex organic compounds present in the dead remains of plants and animals into simple forms. Thus they help in mineral recycling .Dead remains of plants and animals contain inorganic elements in a 'locked' form as components of organic compounds which get relieved by the action of microbes during decomposition. They secrete various types of extracellular enzymes which convert complex organic compounds into simple inorganic compounds which are added into the soil and are vital for plant growth.

2. Sewage Disposal

Ability of microbes mainly bacteria and fungi to decompose organic compounds is of great use in the sanitary disposal of water carried wastes from house hold and industries in the sewage disposal plants. In the sewage disposal plants the solid organic matter is decomposed into liquid sludge that passes through the filters. The filtered water is allowed to pass on to a nearby streams or rivers and the suspended particulate matter is separated and is used as manure.

3. Soil Fertility

Microorganisms help in increasing the fertility of soil. Saprophytic bacteria and fungi convert organic wastes into humus and manure.

Besides it the nitrogen content of the soil is enriched by several nitrogen fixing bacteria and cyanobacteria which fix the atmospheric nitrogen into nitrogenous compound like ammonia which are added into soil for the use of crop plants. Among them most important is *Rhizobium* a symbiotic bacterium present in the root nodules of leguminous plants like gram, groundnut, and soybean. *Azotobacter* (aerobe) and *Clostridium* (anaerobe) are free living nitrogen fixing soil bacteria. Similarly there are number of nitrogen fixing cyanobacteria (blue green algae) like *Nostoc, Anabaena* and *Spirulina* which frequently occurs in paddy fields and make the soil fertile.

Ammonifying bacteria like *Bacillus* and *Sarcina* decompose proteins present in dead remains of plants and animals into ammonium compounds. Nitrifying bacteria like *Nitrosomonas* converts ammonium compounds into nitrites and *Nitrobacter* converts nitrites into nitrates; the chief form which is absorbed by the plants and is vital for their growth.

4. Mycorrhiza

Mycorrhiza is a symbiotic relationship between roots of higher plants and fungi. There are a number of fungal species belonging to *Penicillium, Aspergillus, Cephalosporium* etc. which are present either on the surface or outer tissue of roots getting food which is chiefly the root exudates and the fungi in turn help the plants to absorb minerals from the soil. *Trichoderma* is another common mycorrhizal fungus which checks the growth of pathogenic microbes and hence provides resistance to infection. It has been observed if roots of pines are devoid of mycorrhizal fungi these trees remains stunted in growth.

5. Food and Food Processing

Some of the microbes are used either as food, food supplement or in food processing. An important edible alga is *Chlorella*, a unicellular green alga, which is a rich source of proteins, vitamins, and minerals. Yeast cake which is a nutrient rich food item is prepared by mixing yeast and starch and then it is compressed. It is a good source of vitamin B. *Spirulina*, a blue green alga, contains 60 percent proteins. It is cultured in small tanks and is used as an energy rich food supplement.

A lactic acid bacterium, the *Lactobacillus* is employed in the conversion of milk into curd; a common house hold food commodity. For the conversion of milk into curd first milk is boiled then it is cooled to a temperature of about 30^0C to 35^0C subsequently a spoonful of curd containing lots of lactic acid bacteria is added into 3 to 4 liters of milk in a suitable container that is covered properly. These bacteria multiply and synthesize lactic acid from lactose, the milk sugar and this process is called lactic acid fermentation. Lactic acid coagulates casein, the milk protein, and subsequently milk is converted into curd in about 4 to 5 hours.

6. Industrial Uses

Microorganisms are utilized to produce many commodities of commercial use. Some of their important industrial uses are:

(*i*) ***Alcoholic Fermentation.*** Yeast (*Saccharomyces*) which is a unicellular fungus brings about alcoholic fermentation. It is a process in which anaerobic break down of sugars into alcohol and carbon dioxide takes place with the help of yeast as it has complex enzyme zymogens in it. This process is the basis of two type of industries; brewing and baking industries. Alcohol is used in making various alcoholic beverages like wine, beer, whisky, brandy etc. Carbon dioxide produced in fermentation helps in raising the dough so that baking of cakes, bread, biscuits etc. becomes easier.

(*ii*) ***Curing of Tea and Tobacco.*** Tea leaves after plucking are properly processed to remove water, tannins and other unwanted component and their 'ripening' is done under controlled environmental conditions in which they are subjected to ferment with the help of saprophytic bacteria. This imparts specific colour and aroma to the tea leaves. In the similar fashion curing of tobacco leaves is done.

(*iii*) ***Retting of Jute, Hemp and Flax.*** Fibers from jute, hemp, flax etc. are obtained by a process called retting. In this process after harvesting, the stems of these plants are tied in bundles and are immersed in water for about three weeks. During this period by the action of saprophytic bacteria the fibers of these plants are loosened as the bacteria decompose the soft tissue which binds the fibers firmly with the stem and then their removal become easier.

(*iv*) ***Dairy Industry.*** Dairy industry is related to trading of milk and its products like cheese, curd, butter, 'desi ghee' and 'lassi'. There is a specific role of some microbes in it.

Pencillium, a fungus, is used in processing of cheese; it imparts characteristic flavor and texture to cheese.

To obtain butter and 'desi ghee'; milk is first converted into curd with the help of lactic acid bacteria as explained in a previous section (food and food processing). Butter is extracted from curd by appropriate technique both in houses and dairies and the left over part is called 'lassi'. 'Desi ghee' is prepared from butter.

(*v*) ***Production of Vinegar, Enzymes and Hormones.*** Microbes are used to produce various organic compounds of commercial use. Vinegar (acetic acid) is obtained from molasses and sugars by the action of bacterium *Acetobacter aceti.*

Similarly enzymes like protease, cellulase, pectinase, amylase, and catalase etc. are obtained by the action of fungi lik *Rhizopus japonicas, Aspergillus niger, Pencillium chrysogenum* etc. on appropriate compounds.

Plant hormones like gibberellin are obtained by the action of a fungus *Gibberella fujikuori* which grows on paddy plant and cause 'foolish seedling' disease.

(*vi*) ***Leather Tanning.*** In tanneries the skin of animals are subjected to ferment with microbes chiefly the bacteria which removes soft dermis tissue of the skin.

7. Medicine

Microorganisms are a potent source of different life saving drugs; antibiotics and vaccines. Antibiotics are the chemical substances secreted by one type of microorganisms which kill other type of microorganisms. There are hundreds of antibiotics which are used to cure different diseases of microbial origin. Antibiotic penicillin is obtained from mould fungus *Penicillium notatum* and *P. chrysogenum*. Similarly antibiotics streptomycin, chloromycetin and terramycin are obtained from filamentous bacteria *Streptomyces grieseus*, *S. venezuelae* and *S. aureofaciens* respectively. A fungus *Claviceps purpurea* that causes ergot of rye (*Secale cereale*), a cereal contains many alkaloids like methylergometrine, ergotomine etc. which are used to prepare a drug LSD (Lysergic Acid Diethylamide), which is used in experimental psychiatry. Drugs obtained from ergot are also employed in inducing uterine contraction during child birth.It also regulates uterine haemorrhage during child birth.

Microorganisms are also used to produce vaccines. Technique of vaccination was first of all developed by **Edward Jenner** in 1776 to cure small pox. Vaccines contain attenuated (weakened or killed) strain of pathogenic microbes (bacterium or virus). When a vaccine is injected in the body it activates our blood immune system and protects us from the attack of a specific type of pathogenic organism. For example a vaccine of cholera contains weakened or killed cholera bacteria, and its administration in the body provides protection against cholera and this vaccine was first developed by Pasteur. There are various types of vaccines like vaccine for polio, diphtheria, pertusis (whooping cough) and tetanus which are used to immunize babies at an early age. Small pox has already been eradicated by the use of small pox vaccine.

8. Role of Bacteria and Viruses in Genetic Engineering

Bacteria and viruses are widely employed in genetic engineering or recombinant DNA technology. In genetic engineering a piece of desired segment of DNA (gene) from a donor organism is transferred to cell of other organisms with the help of vectors. Bacterial plasmid and phage DNA are used as 'vectors'. When a vector transfers the foreign gene into a host cell it becomes a part of the host DNA. When such transformed cell replicates the introduced desired gene also multiplies and expresses itself. Crop plants like Bt cotton resistant to pest infection and golden rice rich in vitamin-A contents are some of the transgenic crops formed by recombinant DNA technology using bacteria as vectors.

9. Biogas

Biogas is produced from dung, municipal waste, and other wastes which are mixed with water to form slurry and are digested in a biogas plant. In a biogas plant the organic waste is decomposed by anaerobic bacteria. There are number of naturally occurring saprophytic bacteria that decompose the organic wastes. Production of methane which is the main component of biogas is by the action of archaebacteria the Methanogens. Biogas consists of about 65 percent methane (CH_4), 34 percent carbon dioxide (CO_2) and rest of the other gases like N_2, H_2, H_2S. Biogas is used as a fuel.

HARMFUL ACTIVITIES

Microorganisms have a number of activities which cause immense loss to property as well as life. Some of their negative activities are:

1. Diseases

Microbes cause a number of diseases in plants, animals including human beings.

(i) ***Human Diseases.*** Human suffers with many diseases occurs due to viruses, bacteria, fungi, and protozoa; some of which are fatal. Dengue fever, a fatal disease is caused by Flevivirus and is spread by a mosquito (*Aedes aegypti*). AIDS an incurable disease is caused by HIV (Human Immunodeficiency Virus). Mumps virus, Influenza virus, and Polio virus are the cause of mumps, influenza fever and polio respectively especially in children. Cold which occurs frequently in human is a common viral disease. Bacterial diseases like cholera, tetanus, and typhoid are caused by *Vibrio cholerae*, *Clostridium tetani*, and *Salmonella typhosa* respectively. A common fungal disease 'ring worms' results in the formation of light skin coloured or reddish brown spots on body surface is caused by *Candida* a unicellular yeast like fungus. Malaria is caused by malarial parasite (*Plasmodium vivax)* a protozoan and spread by mosquito (*Anopheles*).Another protozoan *Entamoeba* is a cause of dysentery.

(ii) ***Plant Diseases.*** A number of plant diseases are caused by microbes. Important plant diseases are late blight of potato caused by *Phytophthora infestens* a mould fungus. This disease is recorded in the annals of history of plant diseases as it resulted in Irish famine in 1840 due to failure of potato crop the staple food in Ireland at that time. A million people died due to hunger and another million people were forced to shift to neighbouring countries for their survival. Some other fungal pathogens like *Puccinia* causes rust disease in wheat, *Colletotrichum* causes red rot of sugarcane and *Albugo* causes white rust of crucifers.

Bacterial plant diseases are: citrus canker caused by *Xanthomonas citri*, bacterial leaf blight of paddy caused by *X. oryjae*. Some viral diseases are: mosaic of tobacco caused by TMV (Tobacco Mosaic Virus) and mosaic of bhindi caused by Bhindi Mosaic Virus.

Cephaleuros, a green alga, contrary to other green algae is parasitic in nature that grows on tea plants and causes red rust disease on it.

(iii) ***Animal Diseases.*** Animals do suffer with diseases caused by microbes. Bird flu that occurs in poultry is caused by *Influenza virus-A*. Bird flu when occurs in epidemics takes heavy toll of chickens. Animal cholera caused by cholera bacterium is other reported disease of poultry. Mouth and foot disease of cattle is another serious disease occurs in cows, buffaloes etc. It is a viral disease caused by FMDV (Foot and Mouth Disease Virus).

2. Food Spoilage

Prepared food if remain exposed for long gets spoiled due to the growth of mould fungi like *Mucor, Rhizopus, Penicillium*, and *Aspergillus*. Likewise many saprophytic bacteria like *Bacillus*, and *Staphylococcus* also grow in food. Spoiled food becomes unfit for consumption.

3. Spoilage of Goods and Damage to Buildings

Microorganisms mainly saprophytic bacteria and fungi grow on many house hold items made up of wood and leather especially in rainy season. Bacteria and fungi spoil the house hold items due to their decomposing activities. Likewise, algae grow in the rainy season on buildings and historical monuments year after year loosens the cementing materials, weakens it and also disfigure and damage them if not cared.

4. Food Poisoning

Sometimes prepared food especially milk products when are kept for serving later both in houses as well as in eateries attracts the growth of bacteria like *Clostridium botulinum*. This bacterium secrete a toxin called botulin and contaminate the food, consumption of such contaminated food causes many body disorders and sometimes even becomes fatal. Similarly other bacteria like *Staphylococcus* and some *Bacilli* when grow in stored food secrete toxins and contaminate it.

5. Aflatoxin

Aflatoxin is a toxic substance secreted by fungus *Aspergillus flavus* when grows on exposed grains and seeds especially ground nut seeds. If ground nut seeds are stored in a humid and warm environment Aspergilli grow on it and secret aflatoxin. Aflatoxins are among the most carcinogenic substances known. Children are more prone to aflatoxin and if consumed by them, delay their development and remain stunted in growth.

6. Denitrfication

There are some bacteria that reverse the good done by nitrogen fixing bacteria. These bacteria convert the nitrogenous compounds like nitrites and nitrates present in soil back into nitrogen gas and thus depletes the fertility of soil. This process is called denitrification and such bacteria are called denitrifying bacteria *e.g.*, *Bacillus licheniformis, Pseudomonas aeruginosa.*

CONTROL OF MICROBES

There are various types of pathogenic microbes belonging to viruses, bacteria, fungi and protozoan which cause diseases in plants and animals including human beings. Besides it microbes also spoils food and other valuable articles. Such microbes must be controlled so to check damaged caused by them. Control of microbes means any agent or substance that reduces, inhibits or eliminates microbes. There are two main methods to control microbes which are physical and chemical control of microbes.

1. Physical Control

Physical control of microbes is done by physical agents like temperature, radiations, osmotic pressure and filters.

(*i*) ***Temperature.*** Microbes are generally killed by subjecting them to a high temperature. When food is heated it kills the microbes present in it. Similarly milk is pasteurized when it is heated to 70°C for15 minutes. Surgical implements are sterilized by putting them in boiling water for about 30 minutes. However, bacterial endospores and protozoan spores are very difficult to destroy. The use of autoclave and pressure cooker (heat combined with pressure) is the most reliable way to destroy them.

Freezing of food and use of refrigerators is another way to check the growth of microbes. Freezing of food generally does not kill the bacteria and other microbes but it does not allow their growth, thus helps in preserving of food.

(ii) ***Radiations.*** Different types of radiations are employed to check the growth of microbes. There are two principal types of radiations: ionization and non ionization. Ionization radiations like X-rays, gamma rays are used for sterilizing pharmaceutical and disposable dental and medical supplies such as plastic syringes, surgical gloves and catheters. Non ionization radiations like UV-rays are used to kill microbes on the surface of objects and in the air. A UV or germicidal lamp is found in hospital rooms, operation rooms and cafeteria.

(iii) ***Osmotic Pressure.*** Some food articles like pickles are preserved by adding hypertonic salt solution in it. Hypertonic solution means a solution which is more concentrated than cell sap of microbial cells. Similarly gems and jellies are preserved due to the addition of hypertonic sugar solution in them which acts as a preservative. Hypertonic solutions do not allow the growth of microbes as due to high osmotic pressure of the concentrated salt or sugar solution. Hypertonic solution causes exosmosis of microbial cells means water of microbial cells is extracted out and they are dehydrated and ultimately die. Due to the same reason honey remains bacteria free for a long period.

(iv) ***Filtration.*** Liquids which are thermolabile means looses their property on heating are made bacteria free by filtering through bacteria free filters. There are various type of filters made up of cellulose acetate, cellulose nitrate, teflon etc. which has pore size of about .2 microns that retains bacteria and other microbes when a liquid is passed through it. Similarly water filters employed in houses make the water microbes free.

2. Chemical Control

A wide variety of chemicals known as antimicrobial agents are used to control microbes and include: disinfectants, antiseptics, antibiotics, sulpha drugs and fungicides.

(i) ***Disinfectants.*** Disinfectants are the chemical agents used to lower the level of microbes present on the surface of inanimate objects such as floors, drainage system, etc. Phenols and other compound containing it are commonly used to clean floors that check the growth of microbes on the floor. Chlorine is used to disinfect water particularly in rainy season. For this purpose chlorine in concentration of 0.2 to 0.4 parts per million (ppm) is used as disinfectant for drinking water. A low concentration of sulphur dioxide is used for sterilizing and preservation of squashes.

(ii) ***Antiseptics.*** Antiseptics are the chemicals used on the living tissue such as cuts in the skin to decrease the microbes present on it. Dettol is a commonly used antiseptic employed to apply in low concentration of about one percent on minor cuts of the skin. Similarly antiseptic creams like soframycin, furacin, etc. are applied on the wounds to check the growth of microbes. Similarly iodoform is used as an antiseptic and is applied on the injured body parts.

(iii) ***Antibiotics.*** Antibiotics are the chemicals secreted by one type of microorganisms that kills other microbes. Penicillin is the first antibiotic, isolated from a mould fungus *Penicillium notatum* in 1929 by **Alexander Fleming.** The term antibiotic was coined by **Waksman** in 1944 and he isolated streptomycin from *Streptomyces,* a filamentous bacterium. Antibiotics are treated as *"wonder drug"* and first showed their potential when thousands of wounded soldiers of World War II were saved with the help of use of antibiotics. There are much type of antibiotics and broadly falls in two categories : bactericidal and bacteriostatic. Bactericidal kills the bacteria where as bacteriostatic inhibits the growth of bacteria. Examples of bactericidal are penicillin and ofloxacin and those of bacteriostatic are tetracycline and chloromphenicol.

(*iv*) ***Sulpha Drugs.*** Sulpha drugs are the synthetic antimicrobial agents and are in use from a long time. Before the discovery of antibiotics these were the frequently used drugs to control infection and are still in use. There are a number of sulpha drugs which are frequently used to fight against disease causing microbes. Some important sulpha drugs are: sulphanilamide and sulphadiazine.

(*v*) ***Fungicides.*** Fungicides are the chemical compounds which are used to inhibit or kill the fungi. Chemically fungicides are thousands of different types of inorganic and organic compounds which are used to combat fungal infections in plants, animals including human beings. Fungicides are more frequently used to check fungal diseases in plants as compared to animals. Inorganic fungicides used, are mainly the compounds of sulphur and copper. Most common fungicide used throughout the world is Bordeaux mixture, which consists of copper sulphate, lime, and water in the ratio of 5: 5: 50 respectively. Besides it elemental sulphur is also used to fight fungal infections. There are a large number of organic compounds which are used as fungicides these are mainly quinones, benzene derivatives and heterocyclic nitrogen compounds.

Exercises

1. What are microorganisms and who discovered them?
2. Define microbiology and write a brief historical account of it.
3. Describe the role of Leeuwenhoek, Louis Pasteur and Robert Koch in the field of microbiology.
4. Name the different groups of microorganisms and give characteristics of archaea and bacteria.
5. Write a note on protists.
6. Describe the beneficial activities of microbes.
7. Describe the harmful activities of microbes.
8. Explain the role of microbes in industry.
9. Write a note on role of bacteria and cyanobacteria in soil fertility.
10. What do you mean by control of microbes and how it is done?
11. Write a note on use of microbes in medicine.

❑❑❑

UNIT 3

- Organic Evolution
- Mendel's Laws of Inheritance
- Cell Division
- Nucleic Acids : The Genetic Material

Chapter 7 ORGANIC EVOLUTION

Life exists only on the planet earth. Age of the earth is about 5.2 billion years and the fossil records suggest that life arose on the earth about 3.9 billion years ago. There is enormous type of diversity in the living world. There are thousands of different type of plants, animals and microorganisms. At present only a fraction of these organisms have been identified and named. Many more are still to be identified and named. Not only this, many life forms that existed in the past and are extinct now, represented only by their fossil records. How these life forms came into existence? Man of all ages tried to answer this question. To ponder over this problem two divergent views emerge out. These are a religious view and a scientific view (doctrine of evolution).

RELIGIOUS VIEW (THEORY OF SPECIAL CREATION)

According to this view God, the super natural power is the creator of all the life forms. Different living organisms were created by Him in the form as they exist in the present time. This view is a purely religious without any scientific backing. Buffon (1707–1788) was the first biologist who discarded the religious view of creation of life. According to him life forms are not of fixed type but are changeable and this lead to shift in the approach of understanding of regions of occurrence diverse life forms on a scientific basis by the contemporary naturalists and biologists alike.

SCIENTIFIC VIEW (DOCTRINE OF EVOLUTION)

After **Buffon,** many biologists become interested in analyzing the occurrence of diverse life forms and the consensus emerged out were in favors doctrine of evolution and the important scientists who contributed in it are **Von Baer, Ernst Haeckle, Lamarck, Darwin, Weisman** and **Hugo de Vries**. Evolution may be defined as a slow and gradual change in the earlier simple life forms as a result of which present day complex life forms have came into existence. Evolution is a continuous process and started since the origin of life.

According to the modern theory of origin of life, the first living creatures were unicellular organisms. These were formed in the oceans by chemogeny method around 3.9 billion years ago and at that time the atmosphere of the earth was a reducing one *i.e.,* it was lacking free oxygen (O_2). During that time simple molecules like methane (CH_4), ammonia (NH_3), water vapours (H_2O) and hydrogen (H_2) present in the earth's atmosphere collided with the help of cosmic forces like

UV – radiations, lightening etc. to generate sugars, fatty acids, amino acids, nucleotides etc. which on further condensation formed biomolecules like DNA, RNA, proteins, carbohydrates, lipids etc. and then by the aggregation of these molecules primitive life forms came into existence in an aquatic environment *i.e.,* the oceans. Early life forms were heterotrophic, anaerobic and unicellular organisms. Diverse type of living organisms *i.e.,* microbes, plants and animals get evolved from the primitive life forms gradually in the due course of time first primitive type then advance type. The early life forms utilized the dissolved inorganic and organic compounds present in the oceans. Later, primitive bacteria came into existence. Some of the bacteria become autotrophs as they develop bacterial chlorophyll. After that cyanobacteria (blue green algae) were formed. These primitive oxygenic photosynthetic organisms were the autotrophs that started adding molecular oxygen into the environment. Thus the earth's environment started turning into oxidizing from an earlier reducing. Simple unicellular green algae were formed which later on gave rise filamentous and other higher green algae. Similarly other type of marine and freshwater algae were formed. From algae, bryophytes which are treated as amphibian plants and shares both aquatic as well as terrestrial habitat came into existence. Subsequently first land plants pteridophytes were formed from bryophyte like ancestors. Finally, higher seeded plants gymnosperms and angiosperms came into existence from pteridophytes like ancestors. Photosynthetic plants made the planet earth habitable for the living organisms as they are autotrophs *i.e.,* synthesize their own food and other life forms depend directly or indirectly on plants as a food source. Not only this, O_2 is utilized by other life forms in the food oxidation for liberation of energy. Excess of O_2 is converted into ozone *i.e.,* present about 15 – 30 km around the earth and protects the living organism from the harmful UV–radiations of the sun. Similarly, and simultaneously animals get evolved from the early life forms. First unicellular animals were formed and they gave rise to invertebrates. From invertebrates simple vertebrates like fishes (aquatic animals) came into existence. Amphibians the first land animals were formed from fish like ancestors. Reptiles came into existence from some amphibians. From reptiles higher animals like birds and mammals were formed. Human beings were evolved from ape like ancestors. It is very difficult to explain how and when transformation of one life form into another life form took place but it is an established fact that living organisms evolved by organic evolution and it is an ongoing continuous process that is supported by a number of evidences.

EVIDENCES IN SUPPORT OF EVOLUTION

There are number of evidences to support organic evolution *i.e.,* the formation of the present day higher and complex living organisms from simple and primitive earlier living organisms.

1. Geological Evidence

The sedimentary rocks are the depositories of dead remains of early life forms in the form of fossils. The petrified fossil samples of earlier plants, animals and microorganisms that are represented in the form of impressions, casts etc. give an insight of the type of organisms lived in the past. In these fossils the original shape of body parts of the organisms lived millions of years ago is preserved as such though the original organic component of these body parts have been replaced particle by particle by minerals. Geologists estimate the age of rocks by radioactive dating technique. In this method spontaneous decay of radioactive elements like ${}_{92}U^{238}$, ${}_{6}C^{14}$ that has been locked in the rocks at the time of their formation is recorded. Radioactive isotopes decay with a characteristic half life period *i.e.*, the time required in which half of its original amount is left. Half life period of $_{92}U^{238}$ and $_{6}C^{14}$ is $4.5x10^9$and 6000 yrs respectively. The sedimentary rock of an early age contains fossil of simple organisms as compared to rock of a late age that contains fossils of complex organisms indicating that evolution is an on going process. From the fossil records it was found that among the vertebrates fishes were formed first followed by amphibians, reptiles, birds and mammals. An excellent example to substantiate that birds have evolved from reptiles is justified by a fossil bird *Archaeopteryx* (Fig. 7.1). It is referred by the zoologist as a *'lizard bird'*

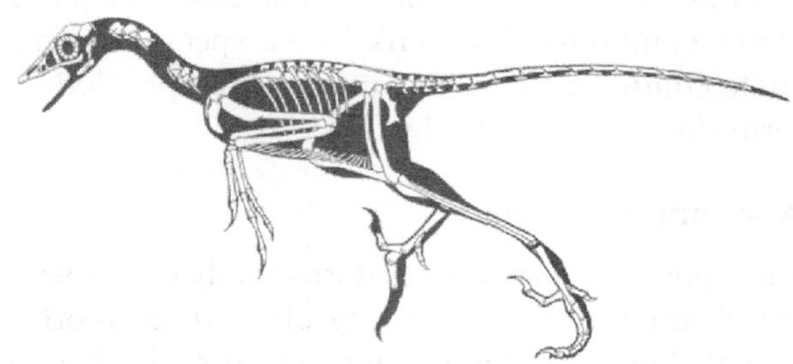

Fig. 7.1. Archaeopteryx – (missing link between reptiles and birds).

as it was having feathers like a bird and teeth in the jaw and claws in the free fingers like a reptile. Such organisms that occurred in the past, present no more now, represented only by fossils and shares characters of two groups of animals are called missing links.

2. Embryological Evidences

The embryological evidences strongly supports the doctrine of organic evolution. The early embryos of the vertebrates *i.e.*, fishes, amphibians, reptiles and mammals are similar in shape and

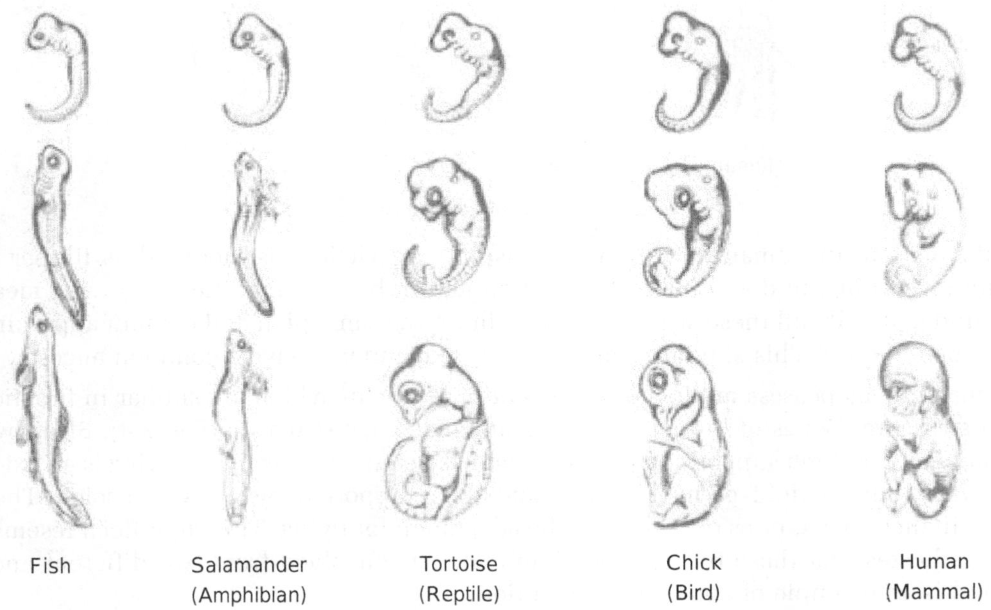

Fig. 7.2. Embryos of vertebrates.

structure. Embryos of all the vertebrates posses gill slits but gills are the respiratory organs only in fishes (aquatic animals) at the adult stage whereas rest of the vertebrates are land animals and have lungs as the respiratory organs instead of gills. Occurrences of gills slits in the early embryonic stages of all the vertebrates suggest a common ancestry of all these animals. On the basis of this observation Ernst Haeckel (1834–1919) proposed the biogenetic law. This law states that ontogeny recapitulates phylogeny which means that during the embryonic development of any organism its complete evolutionary history is repeated. Similarly, seedlings of some plants show resemblances to the plants which may be their ancestors *e.g.*, Australian *Acacia* in its seedlings stage bear bipinnate compound leaves like other species of *Acacia* but adult Australian Acacia do not bear bipinnate compound leaves instead it bear phyllode that is a modification of rachis (petiole) and perform the function of a leaf.

3. Morphological and Anatomical Evidence

A comparative account of morphological features (external body characters) and anatomical feature (internal body characters) of animals provide many clues in support of evolution. There are certain body organs in animals known as **homologous organs** which are similar in structure but differ in functions. Fore limbs of mammals like human, whale, dog and bat (Fig. 7.3) perform

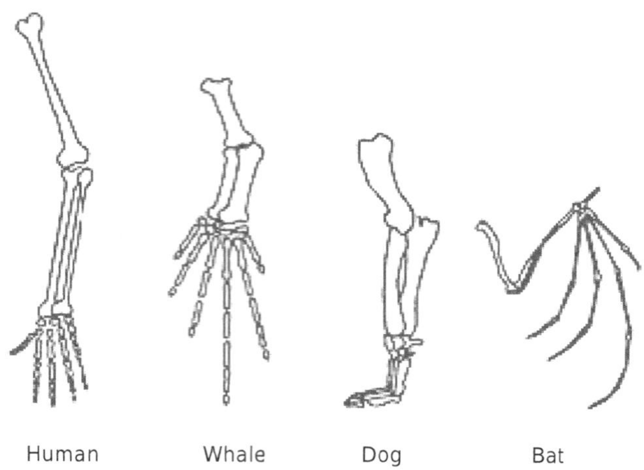

Fig. 7.3. Homologous organs.

different function. In human it is used for grasping, in whale it is modified as flipper and is meant for swimming, in dog it aid in locomotion and in bat it is modified into wing meant for flying. Fundamentally all these structures are built on the same plan and contain approximately similar type of bones. This similarity indicates these mammals have a common ancestry.

Some animals possess **analogous organs** *i.e.*, the organs which are similar in function but differ in structure. Wings of birds and insects are analogous structure (Fig. 7.4). Bird's wing is modification of fore limb supported by muscles and skeleton whereas insects wing is an extension of body integuments (fold of membranes) and it is supported by a few muscles. The only similarity in these two groups of animals is the adaptation for flying. This superficial resemblance in them indicates that these two groups of animals have evolved from two different ancestral stocks and is an example of convergent evolution.

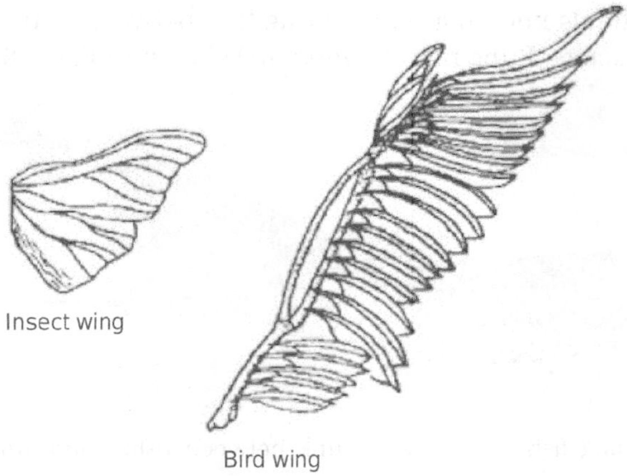

Fig. 7.4. Analogous organs.

A strong refutation to the theory of special creation and support to doctrine of evolution is provided by the **vestigial organs** present in animals. A vestigial organ is the remnant of an organ which at present is of no use to an individual but was well developed in its ancestors. Human body possesses many vestigial organs to name a few these are caudal vertebrae, third eye lid or nictitating membrane, vermiform appendix etc. In humans a rudimentary tail is represented by coccyx which is formed by the fusion of four vertebrae at the end of the vertebral column. Similarly the nictitating membrane that is the third eyelid which is functional in birds and ape and is used for the cleaning of the cornea as it is moved across the eye at intervals. In human being cornea is cleaned by the movement of upper eye lid and hence the nictitating membrane has become redundant and non functional and is present in the form of a small triangular pinkish flap in the inner angle of the eye.

Vermiform appendix present as an extension of caecum in the form of a narrow blind tube between ileum and colon in human alimentary canal is a vestigial organ.But it is well developed in herbivores like rabbit , horse and cattle that harbours cellulose digesting bacteria . Retention of vestigial organs in human body (Fig. 7.5) can not be justified by the theory of special creation. There presence in humans indicate that they were well developed in our ancestors from which we have evolved.

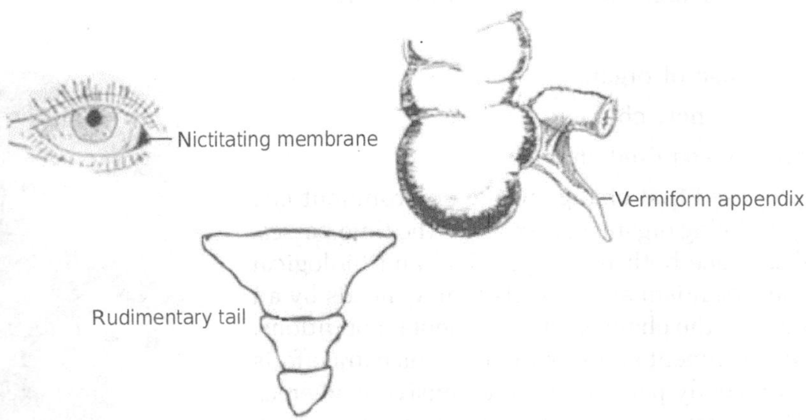

Fig. 7.5. Human vestigial organ.

There are certan animals known as connecting link between two groups of animals as they share some common characters. Lung fish (*Protopterus*) shown in Fig. 7.6 is serves as a connecting

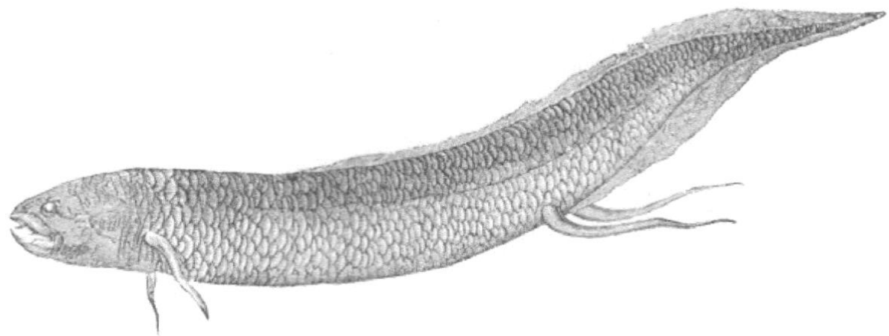

Fig. 7.6. Lung fish - (connecting link between fishes and amphibians).

link between fishes and amphibians.Lung fish besides having all characters of fish also posseses lungs as a respiratory organ as in amphibians indicating evolution of amphibians from fish like ancestors.

THEORIES OF EVOLUTION

The present day species of living organisms have evolved from early life forms by evolution and it is in operation since the origin of life. There are various theories to explain the process of evolution. All the theories of evolution in essence are similar but vary in explaining the manner in which it occurs. They support organic evolution and refute the concept of special creation of life and rigidity of species. According to theories of evolution, species is changeable and new species arise from pre-existing species.

Lamarck's Theory of Inheritance of Acquired Characters

Jean Baptist de Lamarck (1744–1829), a French biologist proposed the theory of inheritance of acquired characters in 1809 in his book *Philospie Zoologique* and explained that new species arises from pre-existing species on account of the interaction between organisms and their environment and the salient points of this theory are:

1. New needs
2. Use and disuse of organs
3. Acquisition of new characters
4. Inheritance of acquired characters.

Fig. 7.7. Lamarck (1744–1923).

Lamarck proposed that change in the environment *i.e.,* the surroundings of living organisms occurs as the time passes. This change takes place both in the physical and biological environment of an organism and it leads to new needs by an organism to cope with the changed environmental conditions. To adopt a new enviornment an individual excerts extra efforts in which some of its body parts are used extensively whereas other body parts are not used in the manner in which they were in use prior to the change. Those body parts which are used extensively by an organism become large and strong and strong and those body parts which disused become weak and small as the time passes. Lamarck in support of his theory gave a classical example of giraffe found in Africa which has a long neck and strong and large shoulders. The ancestors of giraffe were small horse like animals with a small neck. It feeds on tree leaves and to brouse the leaves it has to stretch its neck as a result of which it increased in size and its shoulders became strong and this character is transmitted to the next generation. The present day long necked and strong shouldered girafffe is due to the inheritance of the acquired characters that has been transmitted in its generation after generation. Continuous disuse of an organ makes it weak and small which ultimately becomes redundant. Snakes which are limbless is due to continuous disuse of their limbs.The ancestors of snakes were lizard like reptiles.They were frequently preyed by enemies chiefly the mammals. Out of fear and to protect themselves from enemies they changed their habitat and started living in narrow places such as crevices and under the debris etc. In the new environment they seldom used their limbs for locomotion. Due to continuous disuse of their limbs snakes finally become limbless.

Plants modify themselves according to the environment. The plants growing in dry places develop extensive root system in order to absorb water from the deeper layers of the soil. Leaves of such plants become small and narrow in order to minimise the loss of water by them. Similarly those plants which grow in moist places bear large and broad leaves with shallow root system according to their environment.

Lamarck's theory was of much relevance at the time of its proposition as it disposes the concept of special creation and tried to explain evolution of organisms in a scientific manner. But it was challenged among others by August Wiesmann (1834–1914). He proposed **continuity of germ plasm theory** in 1892. He discarded Lamarck's theory by performing tail mutilation experiment in mouse. He cut the tail of mouse for succesive 22 generation and never obtained a tailless mouse. According to **Weismann's** theory body of a multicellular organism is composed of two type of cells *i.e.*, somatic cells and the germ cells. Somatic cells of an organism contain determinants of body characters in their nuclei. These cells are mortal and thus the characters acquired by an organism during its lifetime are lost when it dies. Whereas the germ cells are immortal and carry the determinants of characters of a species in their nuclei which are transmitted to the next generation.

Darwin's Theory of Natural Selection

The theory of origin of species by natural selection was put forward simultaneously and independently by **Alfred Russel Wallace** and **Charles Darwin** in the year 1858. But the theory given by Darwin got wide spread aclaim across the scientific community as it was well documented. Darwin later in the year 1859 wrote a well known book **origin of species**. In this book he proposed that the present day advanced and complex species of living organisms are the descendants of the early occurring primitive and simple life forms. Prior to the publication of his findings Darwin went on a voyage on the ship H.M.S.Beagle during which he observed the diversity of the flora and fauna of the different regions he visited.He also studied the population theory proposed by **Thomas Rev. Malthus** in 1799 in the form of *"An Essay On Population"* in which he explained that famine, war and diseases inevitably control population growth. Darwin

Fig. 7.8. Darwin (1809–1882).

assembled all these information and formulated the theory of natural selection.This theory is based on the fact that although the living organisms have enormous potential of reproduction yet their population size remains within certain limits as the nature allows only the fit organisms to survive and eliminates the unfit ones. The salient points of this theory are (*i*) over production of offsprings (*ii*) struggle for existence (*iii*) variations (*iv*) elimination of unfavourable variations (*v*) retention of favourable variations and survival of the fittest (*vi*) inheritance of favourable variations and (*vii*) origin of new species.

(*i*) ***Overproduction of Offsprings.*** Every living organism tends to produce much more offsprings than actually survives.A bacterial cell can divide into two bacteria within 30 minutes but they are not allowed to multiply in this manner. An insect lays hundreds of eggs at a time out of which only a few are able to develop into adults.A neem tree produces thousands of seeds each year but none or only a few of its seeds actually grow into trees.

(*ii*) ***Struggle for Existence.*** There is limited availibility of food, space and other natural resources and consequently the organisms compete with each other for the natural resources. Competition is of two main type intraspecific and interspecific. Intraspecific comptetion is among the members of the same species. This type of competition is more severe as their needs are common and they compete for food, space and mate. Interspecific competition is between members of two or more species. This type of competition is less severe as their needs are not so common as in the case of intraspecific competition. They compete for food and other natural resources.

(*iii*) ***Variations.*** Variation is the law of nature. Darwin recognized two main type of variations: continuous variations and discontinuous variations and he considered continuous variations to play a very significant role in evolution and according to him discontinuous variations which occurs all of a sudden and show no gradation, he called them as *'sports'* and treated them as insignificant and ignored their role in evolution. According to him, continuous variations which show various gradations are of prime importance for evolution. No two individuals in this world except for the identical twins are alike. Though members of the same species have many characters in common still they can be easily differentiated from each other. Variations may be favourable or unfavourable. Favourable variations help in the survival chances of an organism and unfavourable variations become a cause of its elimination. During epidemic only some of the individuals of a species who resist that disease survive and others who are susceptible to it die. A slow running rabbit is easily preyed by its enemies as compared to a fast running rabbit. Similarly, strong and healthy members of any type of organisms have an upper edge in life in all respects as compared to weak and diseased ones.

(*iv*) ***Elimination of Unfavourable Variations.*** In the struggle for existence only fit individuals having favourable variations survive and others with unfavourable variations are eliminated. In nature only those individuals who can better adopt in the given environmental conditions are retained and those who can not cope with it are lost.

(*v*) ***Retention of Favourable Variations and Survival of the Fittest.*** In each generation only the fit individuals with favourable variations survive as they can easily adjust themselves according to the environmental conditions and are in a position to obtain all necessary requirements of life. In nature only favourable variations are retained.

(*vi*) ***Inheritance of Favourable Variations.*** Fit individuals who survive transfer the favourable variations to their progenies. This process is repeated in each generation.

(vii) ***Origin of New Species.*** The favourable variations goes on adding generation after generation and over long period of time their cumulative affect results into formation of new life forms. This theory is criticized on the basis that it emphasizes much on variations as the important cause of evolution but did not elaborate the regions of occurrence of variations. Darwin considered only continuous variations and did not consider discontinuous variations and he called them as 'sports' which according to him did not play any role in evolution.

Hugo de Vries' Mutation Theory

Hugo de Vries, a Dutch botanist, proposed the mutation theory of evolution in 1901. Mutations are the sudden heritable changes in genetic matter of living organisms and cause abrupt variations in them. He pointed out that these abrupt variations are important for evolution. He postulated this theory on the basis of his observation on a plant evening primrose (*Oenothera lamarckiana).*He noticed marked variations in the morphological characters of the plants of this species growing wild near his house. When he raised new plants by sowing the seeds obtained from these wild plants he noticed that in the next generation there were some plants with still more varied characters. de Vries explained that these abrupt variations arise as a result of significant change in their genetic matter. According to this theory evolution occurs in a discontinuous and jerky manner rather than gradual and continuous manner as proposed by Darwin. He proposed that new species arise from the pre-existing species in a single generation by a sudden appearance of marked variations.

Fig. 7.9. Hugo de Vries (1848–1935).

Hugo de Vries theory is very significant as he for the first time explained the genetic basis of evolution in living organisms. It is now well established that mutations are a cause of appearance of new characters. But his observation regarding origin of diverse morphological characters in *Oenothera lamarckiana* was later on found to be due to chromosomal aberrations that affect the number of chromosomes and such changes are unstable.

VARIATION AND SPECIATION

Different theories of organic evolutions are of the unanimity that existing species of living organisms arise from pre-existing simple life forms. Each of the theories have explained the process of evolution citing examples in its support but none of the theory has elaborated the concept of evolution fully. The present day concept about evolution is due to genetic variations which act as raw materials for the creation of new species.

Variation

Variations are the law of nature. At present there are millions of different type of living organisms including plants, animals, and microorganisms. Each of which is characterized by certain specific characters. One can easily detect widespread variations among members of a species. No two individuals except the zygotic twins are alike in this world. Variations arise mainly by mutation, recombination and chromosomal aberrations.

1. **Mutation.** Mutation means sudden, discrete and heritable change in the genetic matter of an organism. This term was coined by Hugo de Vries in 1901. Mutations are the most important cause of variations and thus helps in evolution. Mutations are of two types: induced and spontaneous. Mutations occurring in natural conditions are called spontaneous mutations. Now it is an established fact that spontaneous mutations are premeditated *i.e.,* occurs prior to change in environment and not a result of it.. These occur at a very slow rate but their accumulated effect over a long period of time brings a considerable drift in its pre-existing gene pool; modifying it and create new type of genes. Gene pool is a sum total of all genes present in the members of a species. Mutations can also be induced by artificial means by physical mutagens (UV-rays, X-rays etc.) and chemical mutagens (mustard gas, nitric oxide etc.) which are the agents of instantaneous mutations. Mutations add new characters in living organisms and they may be useful or harmful. Useful mutations help organisms to adapt in a new environment conditions.
2. **Recombination.** Recombination means rearrangement of genes and it occurs by crossing over. Crossing over is a process in which exchange of genetic matter between non sister chromatids of homologous pair of chromosomes takes place during pachytene stage of meiosis-I at the time of gamete formation. Recombination brings about reshuffling of genes resulting in the formations of new variants from the existing ones.
3. **Chromosomal Aberrations.** Chromosomal aberrations mean change in the structure and number of chromosomes and it leads to variations in plants and animals.

(i) Structural Changes

Structural modification in chromosomes arise due to deletion, duplication, inversion, or translocation.

(*a*) *Deletion.* It involves cutting of a segment of chromosome which results in loss of some genes.

(*b*) *Duplication.* A segment of chromosome when gets repeated it results in duplication of some genes.

(*c*) *Inversion.* Sometime a segment of chromosome reverses it position. It takes place when a cut segment of chromosome reunites on the same chromosome but it turns to 180° so reverse the location of the genes of that segment.

(*d*) *Translocation.* When a segment of chromosome after detaching gets united with another chromosome.

(ii) Numerical Changes

(*a*) Every living organism is characterized by specific chromosome numbers and most of the higher plants and animals are diploids (having two sets of chromosomes: paternal and maternal). Change in the number of chromosomes is called numerical change and it is of two types: aneuploidy and polyploidy.

(*b*) *Aneuploidy.* Addition or deletion of one or two chromosomes in the fix number of chromosomes of a species is termed as aneuploidy. Loss of one chromosome is termed as monosomy and its gain is called trisomy. In monosomy, one of the homologous pair of chromosome is represented only by a single chromosome whereas in trisomy in addition to a homologous pair there is one extra chromosome. For example 2n number of chromosome in human is 46, but in rare cases there may be 45 chromosome (monosomy) or 47 (trisomy) in some individuals. Such change in number of chromosomes results in characteristic syndromes.

(c) *Polyploidy.* When one or more complete sets of chromosomes are added in the fixed number of chromosomes of a species such changes are termed as polyploidy. It is of two types: autopolyploidy and allopolyploidy. In autopolyploidy there occurs three or more sets of chromosomes which are of the same types. It may be triploid, tetraploid or polyploid which are represented as 3n, 4n, or xn respectively. Allopolyploidy results when set of chromosomes present in a species arise from two or more sources which though originates by cross between closely relate species and such species are fertile. Polyploidy is uncommon in animals but common in plants.

SPECIATION

Speciation means origin of new species from the existing ones. Species is an assemblage of organisms that constitutes its population which are morphologically similar, share common type of characters and its members can interbreed to reproduce fertile offsprings similar to them. New species originates from the pre-existing ones and this process of speciation is slow, gradual and takes considerably long period of time. The modes of formation of new species are mainly allopatric and sympatric speciation, besides it inter specific hybridization and polyploidy also results in formation of new biological species.

Allopatric Speciation. Allopatric speciation refers to formation two or more species from a single species due to fragmentation of original population of a species. When a population is fragmented into two or more small subgroups by geographical changes such as mountain building or separation by long distances such as seas, or deserts as a result of which each of the subgroup is subjected to different environmental stresses and modify accordingly to adjust themselves as per the local needs. Prior to their separation the members of a population were sharing the common gene pool as they were free to interbreed with each other. Due to the geographical separation the members of a subgroup can not interbreed with members of another subgroup as a result of which free gene flow between them is not possible. Each of the subgroup start changing and gradually develops its own gene pool with new characteristics and over a long period of time due to the accumulated effect of the changes in their characters and gene pool a new species start taking its form. Most famous example of allopatric speciation is Darwin's Finches. Darwin during his voyage, before the publication of his book *'Origin of Species'* noticed that the birds of archipelago islands and the birds of mainland *i.e.,* American continent were very different as they were separated long back due to geographical changes and they were not in a position to interbreed and consequently archipelago Finches resulted in the formation of separate species of birds.

Sympatric Speciation. Sympatric speciation occurs due to reproductive isolation. In this mode of speciation members of a population are not isolated geographically, instead due differences in the reproduction timings of its members, they are not allowed to interbreed though inhabiting in the same geographical area. Difference in reproductive timings occurs because of either their dependence on different types of food source or alteration in some of their other habits. Reproductive isolation results in the genetic drift in the original population as a result of which two or more subgroups of original population acquire different sort of gene pools and lead to formation of new species after a considerable long period of time. The most widely accepted example of sympatric speciation is that of chiclid fishes of Lake Nabugabo in East Africa which is thought to be due to sexual selection. In Lake Nabugabo there are many types of chiclid fishes inhibiting in the same lake, though vary as feeds on different preys are supposed to have a common ancestral stock.

Speciation through Hybridization

Hybridization plays a role in speciation and cross between two related species though it is uncommon but such crosses definitely results in creation of new species. Mule a hybrid is formed by mating between male donkey and mare (female horse). Similarly, hinny is a hybrid obtained by mating between stallion (male horse) and female donkey. Both mule and hinny are sterile. In animal husbandry by artificial insemination many high yielding breeds of cow and buffaloes have been created whose milk output is much more as compared to original breeds. Similarly, new breeds of poultry animals have been created by artificial insemination.

Speciation through Polyploidy

Polyploidy is common in plants and allopolyploidy results in formation of new species. About one third of naturally occurring grasses are triploids. Polyploids of many crop plants have been obtained by scientists by duplicating their chromosomes employing suitable techniques. These polyploids have increased the quality as well as productivity of many crop plants. The present day wheat is an allopolyploid of a cross between durum wheat and wild grass that occurred some ten thousand years back.

Exercises

1. What is organic evolution and what are the evidences in support of organic evolution?
2. Write a note on homologous and analogous organs.
3. Write in brief about Lamarck's theory of evolution.
4. What are connecting links and missing links and what is their significance to understand evolution?
5. Name the various theories of organic evolution and describe the Darwin's theory of natural selection.
6. Write a note on mutation theory of evolution.
7. Write a note on genetic variations.
8. What do you mean by speciation? What are the possible modes of speciation?

❑❑❑

Chapter **8**

Mendel's Laws of Inheritance

All the living organisms possess certain easily recognizable characters and reproduce progenies of their own kinds. When the organisms reproduce: the characters are transferred from the parents to the progenies. Reproduction is of two types : asexual and sexual. Asexual reproduction is common in microorganisms, lower animals and all type of plants *e.g.*, a bacterium replicate by binary fission. Some lower animals like *Planaria* has a greater power of regeneration; if it is cut into two or more parts each part regenerates into a young *Planaria*. Likewise, plants like rose, sugarcane, potato etc. are raised by stem cuttings. The progenies produced by asexual reproduction are clones that is the carbon copy of their single parent and resembles it in all respects *e.g.*, rose plants raised by stem cuttings are the clones of the plant from which stem cuttings are obtained. Sexual reproduction occurs in all type of living organisms except prokaryotes. Man is a sexually reproducing organism. It is a common observation that the offsprings *i.e.,* sons and daughter of the same parents are not alike except the zygotic twins. They resemble with each other in many characters but at the same time differ in some other respects. It is also easily detectable that offsprings possess many characters similar to their parents. Sometimes a character that is not present in either of the parents but was present in their ancestors arises in their offsprings. This phenomenon is known as *atavism i.e.,* sudden reappearance of an ancestral character.

How offsprings inherit characters from their parents? How the offsprings of the same parents differs with each other? How characters vary in the members of a species? Genetics is the branch of science that deals with all such problems. It is the study of heredity and variations. Heredity means transference of characters from parents to the offsprings and variation is study of understanding the cause of differences in the characters of offsprings of the same parents as well as differences in the characters of individuals of members of the same species *i.e.,* in a population. It was **George Mendel,** who first of all, gave a scientific explanation of heredity and variation.

Fig. 8.1. George Mendel (1822–1884)

Mendel was born in a peasant family in 1822, at a place Brunn, Austria (now Brono in Czechoslovakia). He eventually became a priest in a monastery at Brunn. He conducted breeding experiments with garden pea (*Pisum sativum*) for about 9 years in the garden of the monastery at Brunn. He was trained in mathematics and science in the childhood and it enabled him to analyze data obtained by him of breeding experiments on the pea plant. He published his findings in 1866 in the journal of local Natural History Society. He died in 1884. Unfortunately no body took notice of this pioneer work in the field of genetics during his life time. However, in the year 1900 three scientists working independently at different places namely **Hugo de Vries** of Netherland, **Carl Correns** of Germany and **Erichvon Tshermak** of Austria while doing breeding experiments came to know about this important work; they recognized its importance, these three scientists are known as re discoverer of Mendel's work. Now Mendel is recognized as *father of genetics*.

TERMINOLOGY USED IN GENETICS

Gene: Gene is a unit of heredity which controls a specific character of an organism. However Mendel used the term factor instead of gene. Chemically gene is a segment of DNA consisting of some specific base sequences and is a determinant of a specific character.

Allele: Alleles are a pair of genes representing the two alternative forms of a character and occupy the same locus on the homologous chromosome. Alleles are represented using appropriate alphabets *e.g.,* T and t are the representation of two alleles of the same gene that controls the plant height and represents tallness and dwarfness respectively.

Homozygous: When two alleles of a character are of same type this is known as homozygous state *e.g.,* TT is the representation of a homozygous tall plant and tt is the representation of a homozygous dwarf plant.

Heterozygous: When two alleles of a character are different *e.g.,* Tt is the representation of a heterozygous tall plant.

Genotype: It means genetic make of an organism *e.g.,* TT and Tt are the genotype of homozygous and heterozygous tall plants respectively.

Phenotype: An individual's physical appearance is designated as phenotype *e.g.,* tall and dwarf plants are the two types of phenotypes representing the plant height.

Dominant: An allele that influences the appearance of a phenotype and do not allow the expression of its recessive counterpart.

Recessive: An allele which is not able to express itself in the presence of its alternative dominant allele. It expresses itself only in the presence of another identical allele.

Mendel formulated the principles of inheritance by experimenting with garden pea. He selected garden pea plant for his breeding experiments because of the following regions:

1. It is a small sized plant, easy to grow, bears bisexual flowers and completes its life cycle in one season.
2. There were different types of the variants (cultivars) of pea plants with contrasting characters such as tall and dwarf plants, purple flowered and white flowered plants etc.
3. These plants were growing in the garden of the monastery in which Mendel was a priest.
4. Garden pea is a self pollinated plant under natural conditions as its flowers completely enclose the reproductive organs (stamens and carpel) till fertilization thus ensuring self pollination. Stamen bears anthers in which pollen grains are formed which eventually forms male gametes and carpel bears embryo sac which contains eggs, the female gametes.

In self pollination pollens from a flower are dusted naturally on the stigma of the carpel of the same flower which ensures fusion of male and female gametes from the same plant.

5. At the same time by artificial means, cross pollination can also be done in garden pea. In cross pollination, pollens from flowers of one plant are dusted onto the stigmas of flowers of other plant. So fusion of male and female gametes from two different plants can be achieved by cross pollination.
6. Hybrids of garden pea are easily obtained by cross fertilization.

Mendel noted in the garden pea seven characters which always appeared in contrasting conditions. These characters are shown in table 8.1. Mendel postulated that these characters are determined by certain discrete units called factors (now known as genes) and each of the genes consists of two alternating forms known as *alleles*. For example, for plant height there is an allele for tallness and another allele for dwarfness. An individual receives one of the alleles of a character from each of the parent.

Table 8.1. Contrasting characters of garden pea studied by Mendel

S. No.	*Character studied*	*Dominant*	*Recessive*
1.	Plant height	Tall(2m)	Dwarf(.5m)
2.	Flower colour	Purple	White
3.	Flower position	Axial	Terminal
4.	Seed colour	Yellow	Green
5.	Seed shape	Round	Wrinkled
6.	Pod shape	Inflated	Constricted
7.	Pod colour	Green	Yellow

First of all Mendel obtained pure lines of 14 cultivars of garden pea *i.e.*, two cultivars each of the 7 contrasting characters by selfing (table 8.1). Pure line plant is a true breeding plant *i.e.*, on selfing (self fertilization) it produces progenies similar to it. For example, a tall plant on selfing produces only tall plants; it is a true breeding or pure line tall plant. In contrast, a hybrid tall pea plant is not a true breeding plant, on selfing it produces two types of progenies one type similar to it *i.e.*, tall pea plants and the other different from it *i.e.*, dwarf pea plants. To obtain pure line tall plants, Mendel raised pea plants by showing their seeds. Among them some were tall and others dwarf. He discarded the dwarf plants and selected only tall plants. Then he obtained the seeds of these selected tall plants allowing selfing. These seeds were sown and the new plants emerged among them he again discarded dwarf plants choose only tall plants. He repeated the same exercise time and again till he got satisfied that on selfing these plants produces only tall plants. Such plants were true breeding or pure line of tall plants. In the similar manner Mendel obtained pure line of rest of the 13 cultivars. Mendel performed many experiments with the help of pure line true breeding plants and devised certain postulates popularly known as *Mendel's laws of inheritance* on the basis of different type of crosses like monohybrid cross, dihybrid cross etc.

Monohybrid Cross

The simplest of the cross conducted by Mendel, called *monohybrid cross* in which pattern of inheritance of one character or trait was taken into consideration For example, a cross between pure line white flowered plants and pure line violet flowered plants (Fig. 8.2).

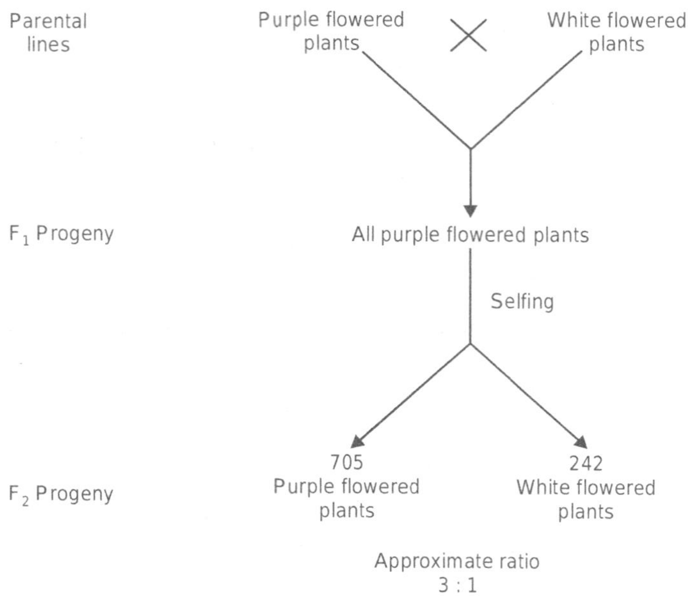

Fig. 8.2. Result of a monohybrid cross in garden pea plant.

In this cross-stigmas of white flowered plants were dusted with pollens of purple flowered plants to cross fertilize it; and it resulted in the formation of a hybrid progeny of first filial generation (F_1 generation). Mendel found that all these plants were purple flowered and resembling with only one of the parents. During this generation there was no trace of the character of the other parent.

Thus in a cross between a true breeding white flowered plant and a true breeding violet flowered plant produced only violet flowered plants in F_1 progeny. When the F_1 generation *i.e.*, purple flowered plants were allowed to self pollinate, the progeny of F_2 obtained was consisted of both purple flowered plants and white flowered plants. Mendel observed that in a total of 929 plants obtained in F_2 progeny there were 705 purple flowered plants and 224 white flowered plants in an approximate ratio of 3: 1. Similar types of results with approximate ratio 3: 1 were obtained for contrasting form of other six traits. All of these results are summarized in table 8.2.

Table 8.2. Results of monohybrid crosses in garden pea as observed by Mendel

S.No.	*Contrasting forms parent's traits*	*F_1 progeny*	*F_2 progeny*	*F2 ratio*
1.	Flower colour (purple × white)	all purple	705 purple 224 white	3.15: 1
2.	Seed shape (round × wrinkled)	all round	5,474 round 1850 wrinkled	2.96: 1

3.	Seed colour (yellow × green)	all yellow	6,022 yellow 2,001 green	3.01: 1
4.	Pod colour (green × yellow)	all green	428 green 152 yellow	2.82: 1
5.	Flower position (axial × terminal)	all axial	651 axial 207 terminal	3.14: 1
6.	Pod shape (inflated × constricted)	all inflated	882 inflated 299 constricted	2.95: 1
7.	Plant height (tall × dwarf)	all tall	787 tall 277 dwarf	2.84: 1

Results

In table 8.2 results of monohybrid crosses of all the seven characters of garden pea studied by Mendel are summarized. From these results following conclusion can be drawn:

(*a*) The F_1progeny of a monohybrid cross of garden pea plant for any of the character all plants exhibit only one form of the character. For example on the basis of flower colour there are two cultivars of pea *i.e.,* purple flowered and white flowered plants and a cross between a true breeding purple flowered and a true breeding white flowered pea plant, in F_1 generation only purple flowered plants are obtained and trait for white colour remains hidden. The trait which is expressed is called *dominant character* and the trait which remains hidden is called *recessive character.* Thus purple colour of the flower is a dominant character and its white colour is a recessive character in garden pea. Similarly a cross between true breeding tall plants and true breeding dwarf plants resulted in the formation of all tall plants in F_1 progeny so tall character is dominant over dwarf character.

(*b*) Results of the reciprocal crosses are same *i.e.,* in a cross between homozygous tall and a homozygous dwarf plant pollens can be obtained from a dwarf plant and are dusted on the stigma of a tall plant or vice versa *i.e.,* the source of pollens is a tall plant and they are dusted on the stigma of a dwarf plant in both the cases the result will be same *i.e.,* only tall plants are formed in F_1 progeny.

(*c*) In F_2 progeny which is obtained by selfing(self pollination) of F_1 plants both of the parental forms of traits are expressed in the approximate ratio of 3: 1.

On the basis of cross breeding experiments Mendel proposed certain postulates which are generally referred as laws of inheritance which are: (*i*) law of dominance, (*ii*) law of segregation and purity of gametes and (*iii*) law of independent assortment.

(i) Law of Dominance

As shown in a monohybrid cross between pure lines purple flowered and white flowered plants in F_1 progeny only purple flowered plants are formed (Fig. 8.2). The trait which expresses itself is called the dominant trait as compared to the trait which is unable to express itself and is called recessive trait. So the dominant trait in the flower colour is purple and recessive trait is white. Thus according to the law of dominance when homozygous organisms (pure line organisms) differing in single pair of contrasting traits of characters, are crossed with each other in F_1 progeny out of the two traits of a character only one trait is expressed. The results of monohybrid crosses, of all the seven characters, studied by Mendel showing the dominant traits in F_1 progeny are depicted in table 8.2 and the dominant and recessive characters in garden pea studied by Mendel are shown in table 8.1.

(ii) Law of Segregation and Purity of Gametes

The law of segregation states that during the gamete formation two alleles of a gene separate, so that each gamete receives only one allele out of the two. To understand the law of segregation and purity of gametes consider a cross between two parents, a homozygous tall and a homozygous dwarf garden pea plants as shown in Fig. 8.3. In this cross source of sperms (male gametes) is a tall parent of the genotype (TT) which at the time of gamete formation forms only one type of male gamete designated as (T) due to segregation of the parental alleles. The other parent is a dwarf plant of the genotype (tt) which is the source of eggs(female gamete) and during gamete formation it forms one type of female gametes designated as (t) due to segregation of parental alleles. In this cross the plants of F_1 progeny all are tall (heterozygous) having the genotype (Tt) formed by the fusion of sperm (T) and egg (t). In this generation there is no trace of dwarf

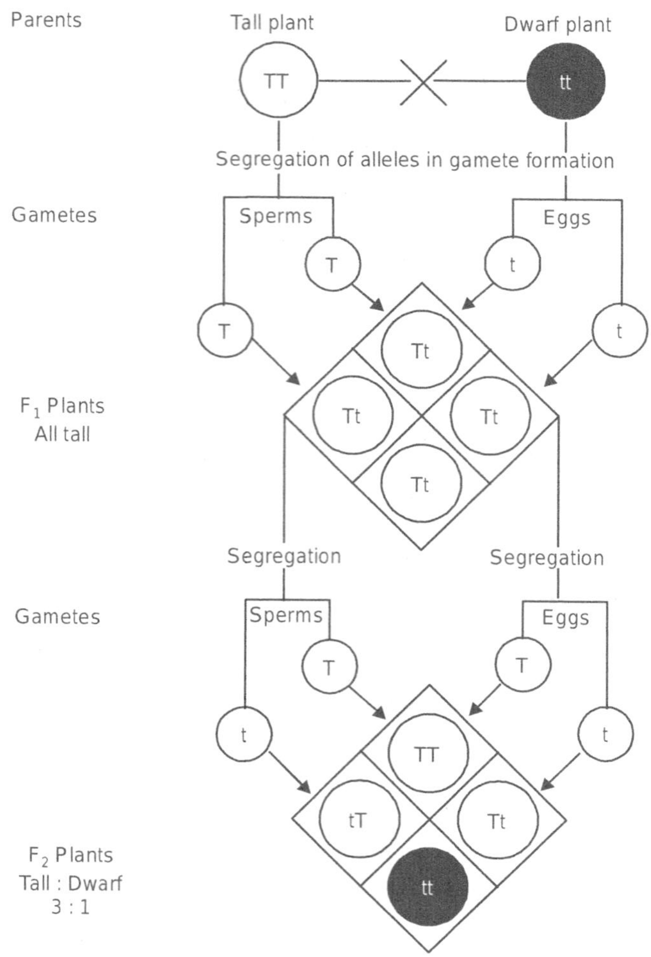

Fig. 8.3. Cross between tall and dwarf pea plant showing segregation of alleles.

character. On self pollination F_1 plants give rise to F_2 progeny which consists of both tall and dwarf plants in the ratio of 3: 1. This shows that the allele controlling the dwarf character maintains its presence in F_1 progeny and is not lost though it can not express itself. The reappearance of dwarf plants in F_2 progeny is due to the probability factor when male and female gamete each of which has an allele (t) fuses resulting in the formation of a dwarf plant of the genotype tt. This ensures purity of gametes *i.e.*, an allele maintains its characteristics or its property is not spoiled or altered in the presence of its dominant counterpart.

(iii) Law of Independent Assortment

This law states that alleles of different genes assort independently of each other during gamete formation as well as their expression in the subsequent generation. When a cross is done between parents differing in two or more pairs of contrasting characters, the distribution of factors(genes) of each character in the gametes is independent of each other and they also expresses themselves in subsequent generation independent of each other. This law can be explained with the help of a dihybrid cross.

Dihybrid Cross

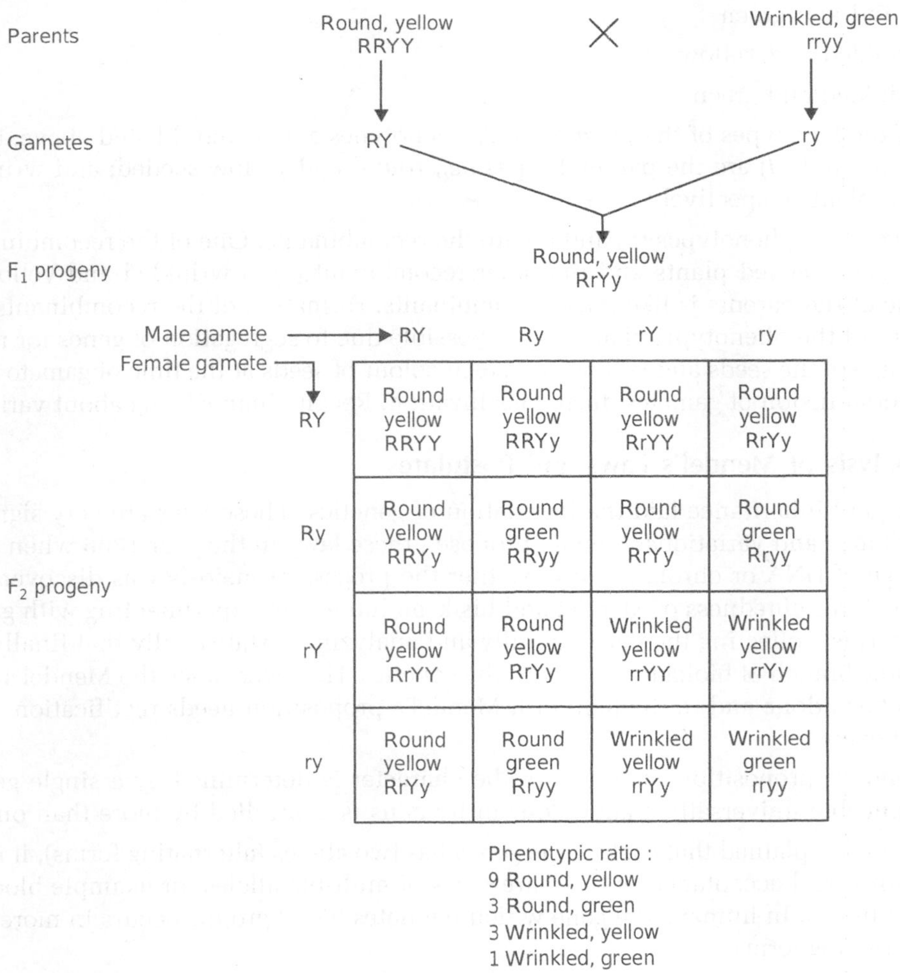

Fig. 8.4. A dihybrid cross showing independent assortment of genes.

With the help of a dihybrid cross pattern of inheritance of two characters is studied at a time. One of the dihybrid cross done by Mendel was a cross between a homozygous round and yellow seeded pea plant (RRYY) with another homozygous wrinkled and green seeded (rryy) pea plant Fig. 8.4. The two of the characters of garden pea studied in this cross are shape and colour of the seeds. R and Y stand for round and yellow seeds respectively whereas r and y stands for wrinkled and green seeds respectively.

In the pea plant round character of seed shape is dominant over its wrinkled shape whereas the character yellow colour of seed is dominant over green colour. During gamete formation the parent having genotype RRYY produces gametes composed of RY alleles and the parent with rryy genotype produces gametes with ry alleles. The gametes RY and ry fuses to form a hybrids with a genotype RrYy(heterozygous) in F_1 progeny. All plants of this generation are phenotypically similar *i.e.*, round and yellow seeded plants.

Selfing is done in F_1 plants. During gametogenesis four types of gametes viz. RY, Ry, rY and ry are formed in both of the parents and on fertilization four different types of phenotypes are formed in F_2 progeny in the ratio given below:

(*a*) Round and yellow : 9

(*b*) Round and green : 3

(*c*) Wrinkled and yellow : 3

(*d*) Wrinkled and green : 1

Out of the four types of the progeny of F_2 phenotypes a, b, c and d listed above, two of the phenotypes (*a*) and (*d*) are the parental types *i.e.*, round and yellow seeded; and wrinkled and green seeded plants respectively.

The two other phenotypes (*b*) and (*c*) are the recombinants. One of the recombinants (*b*) is round and green seeded plants and the other recombinant (*c*) is wrinkled and yellow seeded plants. None of the parents is like these recombinants. Formation of the recombinants *i.e.*, new combinations of the phenotypic characters is possible due to segregation of genes for round and wrinkled shape of the seeds and yellow and green colour of seeds at the time of gamete formation and the random fusion of gametes during fertilization. Recombinants bring about variations.

Critical Analysis of Mendel's Laws and Postulates

Mendel's laws of inheritance laid the foundation of genetics. These laws are very significant to study inheritance and variations. Mendel proposed these laws in the year 1866 when there was no knowledge of DNA or chromosomes. Neither the process of meiosis was discovered at that time. It is the foresightedness of Mendel and his keen interest in experimenting with garden pea to obtain hybrids, collecting the data carefully and analyzing it statistically that finally took the shape of a new branch of biological sciences the genetics. However, since the Mendel's time a lot of work has been done and in the light of it Mendel's proposition needs rectification which are summarized below:

1. Mendel's proposition that each of the character is determined by a single gene is not applicable universally. Skin colour in humans is controlled by more than one gene.
2. Mendel explained that each of the genes has two alleles (alternating forms), it is also not of universal acceptance as there are cases of multiple alleles for example blood groups in humans. In humans the gene which regulates blood groups occurs in more than two alternating forms.

3. Law of dominance is not followed regularly as in *Mirabilis jalapa*, a common flowering plant. In this plant there are two types of plants, one type of it bears red flowers and the other type has white flowers and when a cross is done between two of such plants gives a hybrid bearing pink flowers. This shows semi dominance.
4. Law of segregation and purity of gametes is the only law of Mendel's which is of universal occurrence. It is applicable to each and every sexually reproducing organism. According to the law of segregation, two alleles of a character segregate at the time of gamete formation. This law becomes well established when the process of meiosis was discovered. During meiosis there is separation of homologous pair of chromosomes so that each of the gamete receives half the number of chromosomes to that of the parent cell.
5. Law of independent assortment of genes is also not reported in all the cases. This law holds good only if different pairs of genes either occur on different but not homologous chromosomes or if occur on the same chromosome they must lie far away from each other. This law fails in cases of gene interactions in which effect of a gene is influenced (masked or modified) by another non allelic gene. This law also fails in those cases in which linkage is reported. Linkage means tendency of inheriting together of all the genes lying together on a chromosome during gamete formation so they are expressed simultaneously in the progeny.

Exercises

1. Who is known as father of genetics?
2. On which plant Mendel did work?
3. Why Mendel choose garden pea plant for his research work?
4. Explain the terms genetics, gene and allele.
5. Name the scientists who are known as rediscoverers of Mendel's work.
6. Which of the Mendel's law is of universal acceptance and why it is so?
7. Name the different characters of garden pea plant studied by Mendel.
8. Explain the law of dominance with a suitable example.
9. Describe the law of independent assortment of genes.
10. Explain the law of segregation and purity of gametes.
11. Give a critical analysis of Mendel's laws and postulates.

❑❑❑

Chapter 9 CELL DIVISION

New cells arise by the division of pre-existing cells. Cells have a unique property of self duplication. This property of self replication of cells is the basis of continuance of life. In unicellular organisms, cell division aids in the formation of two individuals from the earlier one. A fully grown amoeba divides into two young amoebae by cell division. In a multicellular organism cell division helps both in its growth and reproduction. Life of a multicellular eukaryotic organism starts with a single cell zygote (fertilized egg). Zygote by repeated cell divisions forms trillion of cells present in the body of a mature multicellular organism. These cells get differentiated during the early developmental stages and form various organs present in a mature organism. When an organism grows, new body cells are added in it, by the division of some of its actively dividing cells. However, in mammals, nerve cells never divide after birth. Cell division continues throughout the life of an individual. In a mature organism not all of its cells divide but only the cells of some regions divide. In human body, cells of bone marrow and germinal epithelium continue to divide; and it occurs till an individual survives. Similarly in mature plants, cells of meristematic regions, present at their apical tips *i.e.,* at root apex and shoot apex continue to divide throughout their life.

Cell division is of two type: mitosis and meiosis. Mitosis occurs in the somatic cells (body cells) and meiosis takes place in the gonidial cell (germ cells). An actively dividing somatic cell enters into a cell cycle.

CELL CYCLE

Each somatic cell is capable of division which passes through a cycle called cell cycle (Fig. 9.1). Cell cycle is an orderly sequence of events consisting of a long interphase and a comparatively short M-phase or mitosis. Interphase is the stage during which all necessary arrangements of cell division takes place by duplicating its chromosomes and other cytoplasmic contents. It was earlier referred as *'resting stage'* but contrary to it, this is the stage of intense biosynthetic activity. It consists of three stages G1-phase, S-phase and G2-phase.

(*i*) **G1-Phase or the Post Mitotic Growth.** It is the first stage of interphase in which a cell grows and it synthesizes RNA and proteins required for DNA replication.

(*ii*) **S-Phase or Synthetic Phase.** It is the second stage of interphase in which DNA contents of the dividing cell is duplicated. At the end of this phase the cell has duplicated chromosomes and each of the chromosomes is now composed of two sister chromatids held together by a common centromere.

(*iii*) **G2 Phase.** After S-phase the cell enters into G2-phase or pre-mitotic growth which is the third and last stage of interphase. In this phase cell grows further. The synthesis of RNA and proteins continues but DNA synthesis is stopped.

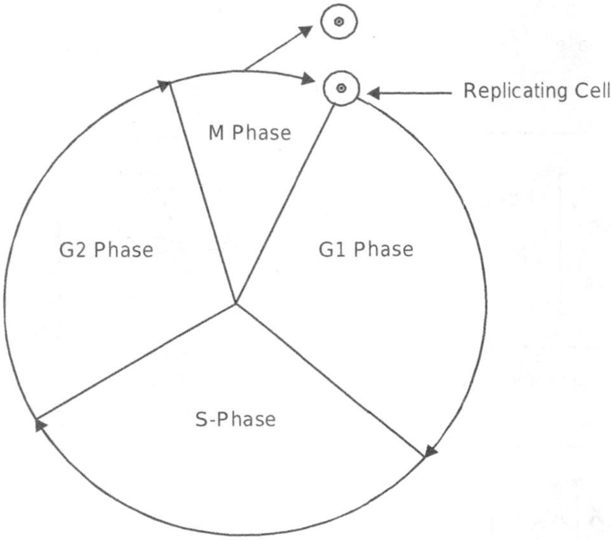

Fig. 9.1. Mitotic cell cycle.

G2-phase is followed by M-phase or mitosis during which the sister chromatids of each duplicated chromosome separates one sister chromatid going to each of the daughter cell.

MITOSIS

It is an equational cell division in which a cell divides into two daughter cells each of which has the same number of chromosomes as that of parent cell. The term *mitosis* was coined by German biologist **Flemming** (1882). In mitosis two important events takes place which are *karyokinesis* (division of nucleus) and *cytokinesis* (division of cytoplasm). It consists of four stages: prophase, metaphase, anaphase and telophase. Various stages of mitosis are shown in Fig. 9.2.

1. Prophase

It is the first stage of mitosis. During this stage nucleus becomes spheroid and its viscosity increases. Chromosomes which remain in a highly disperse state as fine chromatin fibres during interphase start making their appearance as distinct thread like structures by gradual coiling and condensation of chromatin fibres. Nuclear membrane and nucleolus start disappearing gradually and subsequently disappear. With the disapperance of nuclear membrane, cytoplasm and nucleoplasm intermingle with each other. At the end of prophase or at prometaphase stage the duplicated nature of chromosomes also becomes distinct although it occurs

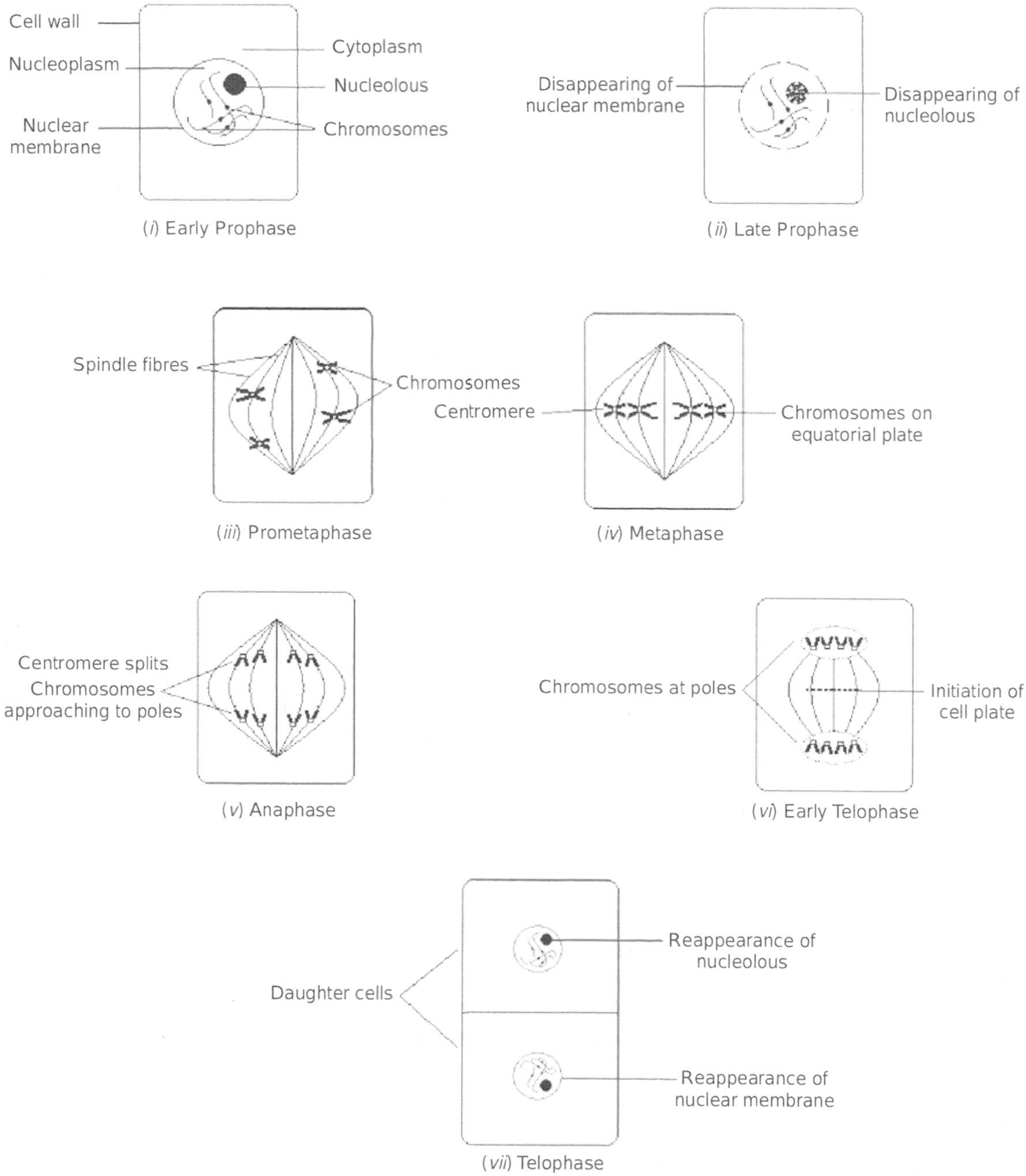

Fig. 9.2. Stages of mitosis in a plant cell.

in interphase stage. Formation of spindle fibres is also initiated and chromosomes start arranging on them. At the end of prophase or prometaphase spindle fibres start making their appearance and the duplicated chromosomes start arranging themselves on the spindle fibres.

2. Metaphase

It is the second stage of mitosis. During this phase chromosomes become most distinct as they are in a highly condensed state. A bipolar spindle apparatus is formed. It consist of fine fibres of protein *tubulin*. Chromosomes get themselves arranged on the equatorial plate of the spindle apparatus. They are arranged in such a way that the common centromere of each of the duplicated chromosome lie on the equatorial line and the arms of its two sister chromatids are directed away from the equatorial line.

3. Anaphase

It is the third stage of mitosis. This phase is initiated by the splitting of common centromeres of each of the duplicated chromosome. It results in the formation of two chromosomes (formerly two sister chromatids) from each of the duplicated chromosome. Consequently two identical sets of the chromosomes are formed. These two sets of the chromosomes are pulled towards the opposite poles. Chromosomes are pulled towards the opposite poles as their centromeres are attached with the spindle fibres and these fibres contract gradually pulling them towards the opposite poles. Finally the two sets of the chromosomes reaches to the opposite poles.

4. Telophase

It is fourth and the last stage of mitosis. It is initiated when two sets of the chromosomes reaches to the opposite poles of cell. The changes which occur in this stage are reverse to that of the prophase stage. Nuclear membrane starts appearing around each set of the chromosomes and gradually two nuclei are organized at the opposite poles of the cell. Nucleolus also appears in each of the nucleus. Chromosomes in each of the nucleus start uncoiling and gradually their thinning occurs.

With the formation of two daughter nuclei the process of nuclear division *i.e.,* karyokinesis is completed. Simultaneously, division of cytoplasm *i.e.,* cytokinesis also takes place. In plant cells it is initiated by the formation of a cell plate at the centre of the dividing cell which gradually extends laterally and ultimately divides the cell into two daughter cells. However, in animal cells cytokinesis occurs by the formation of a cleavage furrow in the dividing cell. During that plasma membrane is constricted in the middle of the dividing cell which gradually deepens and ultimately results in the formation of two daughter cells.

Significance of Mitosis

Mitosis is of great significance to the living organisms and is absolutely essential for their growth, regeneration, repair and maintenance. All the cells present in the body of a multicellular organism are genetically alike and are formed by the repeated mitotic cell divisions and their origin can be traced to a single cell *i.e.,* zygote. Mitosis helps in replacing the old, worn out and damaged cells by forming new cells in place of them. Similarly it also helps in the healing of wounds.

MEIOSIS

It is a reductional cell division in which a diploid cell divides by two successive divisions (Meiosis-I and Meiosis-II) resulting into the formation of four haploid cells each of which have half the number of chromosomes to that of the parent cell. Different stages of meiosis in an animal cell are shown in Fig. 9.3. In higher organisms it occurs only in the gonidial cells or the germs cells at the time of gamete formation. The term *meiosis* was coined by **G.B.Farmer** and **G.E.Moore** (1905).

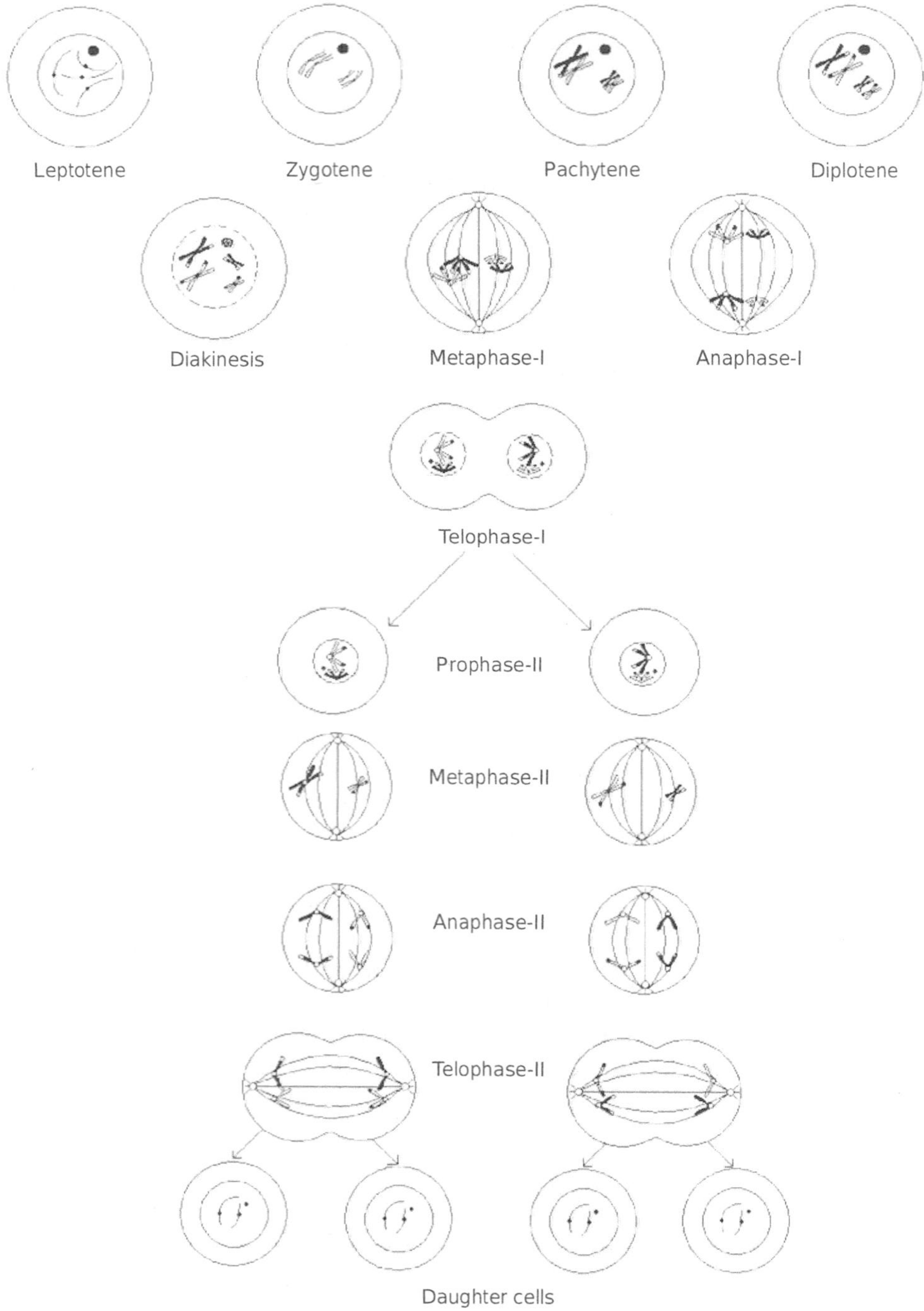

Fig. 9.3. Stages of meiosis in an animal cell.

Meiosis-I

In this division chromosome number is halved and it is preceded by an interphase comprising of first growth phase G1, synthetic phase S and second growth phase G2 similar to mitosis. Meiosis-I is divided into four stages *i.e.,* Prophase–I, Metaphase–I, Anaphase-I and Telophase-I.

Prophase-I

It is a comparatively long and complex stage and consists of five sub stages (*a*) leptotene, (*b*) zygotene, (*c*) pachytene, (*d*) diplotene and (*e*) diakinesis.

(*a*) ***Leptotene.*** Chromosomes which remain in a dispersed state in the interphase as chromatin fibres start making their appearance as fine threads at the beginning of leptotene by gradual coiling and condensation of the chromatin fibres. Afterwards they become shorter and thicker.

(*b*) ***Zygotene.*** In this stage homologous chromosome comes close to each other. In a diploid cell there are two sets of chromosomes, each of the sets is obtained from either of the parent during sexual reproduction. The homologous chromosomes are similar in length and in the position of their centromeres. In each pair of a homologue there is one paternal chromosome and one maternal chromosome. Pairing of homologous chromosomes is called **synapsis** and a pair of homologous chromosomes is called a bivalent. In a bivalent each homologous chromosome is consists of two closely opposed sister chromatids and hence each of the bivalent has four chromatids which is also called a tetrad.

(*c*) ***Pachytene.*** The chromosomes become thicker and shorter by further coiling and condensation. During this stage crossing over occurs. *Crossing over* is the exchange of fragments of chromatids between non sister chromatids of homologous pair of chromosomes resulting in rearrangements of genes (exchange of genes between non sister chromatids). Non sister chromatids are paternal and maternal chromatids of the homologue. This exchange occurs by breakage and reunion of fragments of non sister chromatids and at the points where the exchange takes place cross like structures called *chiasmata* (singular chiasma) are formed. Pachytene is followed by diplotene.

(*d*) ***Diplotene.*** In diplotene, thickening and shortening of chromosomes continue and this stage is initiated when homologous chromosomes start departing from each other. Gradually they move away from each other. As a consequence of crossing over, taking place in pachytene, the homologous chromosomes now have chromatids with rearrangements of genes in some of their segments. Next stage is diakinesis.

(*e*) ***Diakinesis.*** The homologous chromosomes still move away from each other and they become shorter and thicker. During this stage nuclear membrane and nucleolus which start disappearing at the onset of prophase-I disappear totally. Spindle fibres formation is initiated and the chromosomes start arranging on these fibres. Prophase-I is followed by metaphase-I.

Metaphase-I

Bipolar spindle formation is completed in this stage and bivalents are arranged on the equatorial plate. Chromosomes are the shortest and most distinct during this stage. Homologous chromosomes of a bivalent are arranged on the equator in such a way that their chromatids face each other and lie on the equatorial line whereas their centromeres lie away from each other and towards the opposite poles.

Anaphase-I

This stage is initiated when the homologous chromosomes of each bivalent start pulling towards opposite poles. At this stage each of the chromosomes is consist of two chromatids which are held together by a common centromere. Centromeres of the chromosomes are attached with the spindle fibres and by the gradual contraction of the latter the chromosomes finally reach to opposite poles. As only one chromosome of each bivalent reaches to the opposite poles consequently at the end of anaphase-I each pole receives half the number of chromosomes or haploid number of chromosomes.

Telophase-I

At the onset of telophase-I nuclear membrane start appearing around the chromosomes at each of the poles and finally it results into formation of two daughter nuclei ; each of which has haploid number of chromosomes though each chromosome has two chromatids held together by a common centromere.This is generally followed by cytokinesis which results in the formation of two haploid daughter cells thus completing the meiosis-I.

In between meiosis-I and meiosis-II there may be a short intra meiotic interphase. However, no replication of DNA occurs in this phase. Sometimes although daughter nuclei are not fully organized even at the end of telophase-I yet the dividing cell may direct enter into meiosis-II.

Meiosis-II

It is an equational division similar to mitosis. During meiosis-II each of the two daughter cells, formed at the end of meiosis-I, having haploid number of chromosomes and each of the chromosomes consists of two chromatids held by a common centromere further divides and as a result of which finally four haploid cells are formed at the end of meiosis. Meiosis-II consists of prophase-II, metaphase-II, anaphase-II and telophase-II.

Prophase-II

It is quite simple and of short duration as compared to prophase-I. It is similar to prophase of mitosis. In this stage nuclear membrane and nucleolus disappears. Chromosomes are shortened and thickened. Appearance of spindle fibres is initiated.

Metaphase-II

Bipolar spindle formation is completed and the chromosomes are arranged on the equatorial plate. Chromosomes are arranged in a fashion that their centromeres lies on the equatorial line and the two chromatids of each of the chromosomes lie away from each other towards the opposite poles.

Anaphase-II

At the onset of this phase common centromere of each chromosome splits and the two sister chromatids are separated and they move towards the opposite poles. Each of these chromatids hence forth is called **chromosome**. These chromosomes are pulled towards the opposite poles as their centromeres are attached with spindle fibres and these fibres contracts gradually. Finally each pole receives one haploid set of chromosomes.

Telophase-II

During this stage each set of haploid chromosomes alongwith a nuclear membrane start arising as a result of which a daughter nucleus is organized at each pole. A nucleolus is also reconstituted in each of the nucleus. Chromosomes start uncoiling and become thin.

Finally cytokinesis occurs and two nuclei are separated. Each daughter cell of meiosis - I thus produces two haploid daughter cells at the end of meiosis-II. So in all product of meiosis of a diploid gonidial cell are four haploid daughter cells (gametes).

Significance of Meiosis

Meiosis helps in maintaining the constant number of chromosomes in sexually reproducing organisms. During fertilization male and female gametes have haploid number of chromosomes and they fuse to form a zygote restoring the diploid number of chromosomes. If meiosis does not occur during gamete formation they will be diploid and when a zygote so formed by their fusion will be tetraploid and this duplication will be repeated in each generation resulting into a distorted species whose survival is not possible. Thus meiosis helps in maintaining constant number of chromosomes in a species and each species has a specific number of chromosomes. Meiosis also causes variations due to crossing over as the four daughter cells formed in meiosis differ with each other as well as the parent cell in their genetic constitution. So meiosis results in variation in the genetic character in the offsprings. Variations in the long run also help in evolution.

Table 9.1. Difference between mitosis and meiosis

S.No.	*Mitosis*	*Meiosis*
1.	Mitosis is an equational cell division during which a cell divides into two daughter cells each having the same number of chromosomes as that of parent cell.	Meiosis is a reductional cell division during which a cell divides into four daughter cells each having the half the number of chromosomes to that of parent cell.
2.	It is completed in a comparatively short duration during which a cell divides only once.	It is completed in a comparatively longer duration during which a cell divides by two successive divisions (meiosis-I and meiosis-II).
3.	Prophase stage is simple and takes short time for completion.	Prophase-I is a complex stage, takes more time for completion and consists of five sub stages.
4.	There does not occur pairing of homologous chromosomes during its prophase stage.	During its prophase-I stage pairing of homologous chromosomes as bivalents occur.
5.	No crossing over occurs in it.	Crossing over during which exchange of fragments of chromatids between non sister chromatids of homologous chromosomes takes place occurs in prophase-I.
6.	During its metaphase stage chromosomes are arranged on the equatorial plate of spindle apparatus in such a way that centromere of each chromosome lie on the equatorial line and its sister chromatids lie in opposite direction away from the equatorial line towards the opposite poles.	During metaphase-I stage of meiosis two homologous chromosomes of a bivalent are arranged on the equatorial plate of spindle apparatus in such a way that their chromatids faces each other and lie towards the equatorial line and their centromeres lie away from the equatorial line and towards the opposite poles.

7.	Division of the centromere takes place in anaphase separating two sister chromatids of a chromosome.	There is a delayed centromere division as it does not divide during anaphase-I stage. In this stage homologous chromosome rather than chromatids separate. Centromeres divide during anaphase-II.
8.	Mitosis occurs in somatic cells and the daughter cells are all alike in their genetic constitution similar to the parent cells. It occurs both in haploid as well as diploid cells.	Meiosis occurs only in diploid cells. It brings about variations as the daughter cells are not alike; they differ in their genetic make up with respect to each other as well as that of the parent cells.
9.	Mitosis helps in growth, regeneration, repair and maintenance of living organisms.	Meiosis helps in the formation of gametes in the sexually reproducing multicellular organisms.

Exercises

1. Write a note on cell cycle.
2. Draw only well labeled diagrams of different stages of mitosis.
3. Define mitosis and explain it with the help of diagrams.
4. What is the significance of mitosis?
5. Describe prophase-I of meiosis.
6. Define meiosis and write about its significance.
7. Differentiate between mitosis and meiosis.

❑❑❑

Chapter 10

NUCLEIC ACIDS: THE GENETIC MATERIAL

There are millions of different types of living organisms including plants, animals and microorganisms. Each type of organism is characterized by some specific characters. It is very easy to differentiate a mango tree, an elephant and a human being. Moreover, members of a single species are not alike. For example all human beings are members of the same species *Homo sapiens*; still no two human beings except zygotic twins are identical. Each and every human being is unique and can be easily differentiated from each other on the basis of their appearance and other characters like height, colour of skin and eyes, shape of nose and ears, pitch of sound, intelligence etc. What is the basis of specific characters possessed by different living organisms and the variants found among the members of the same species? The basis of differences in the specific characters of various types of living organisms and the variants present among the individuals of the same species is the genetic matter of living organisms.

Genetic matter is the substance that stores all the information that are expressed in terms of phenotypic characters; the external observable characters and are transmitted from parents to the offsprings generation after generation; and DNA(deoxyribose nucleic acid) is the genetic matter of all living organisms except some viruses like (TMV) tobacco mosaic virus. TMV is composed of only RNA and proteins and its genetic matter is RNA. Bacteriophages and all other prokaryotic and eukaryotic organisms possess DNA. DNA is present in the cells of all living organisms. It is the only self replicating biomolecule of the cell. In eukaryotic cells it is present as nucleoproteins (DNA plus histones) or chromosomes in their nuclei. In prokaryotic organisms like bacteria and cyanobacteria it is present in the form of a long circular and naked DNA without histones in the cytoplasm as prokaryotic cell lacks a nucleus (Chapter 1). In any multicellular organisms all of its cells are genetically alike as they are formed by mitotic cell division and their origin can be traced to a single cell *i.e.*, zygote (Chapter-9).DNA was first discovered by **Mischer** (1887); and its structure was first explained by **Watson** and **Crick** who proposed a double helix model of it in 1953 (Chapter-2) that has been widely accepted and were awarded Noble Prize for this work in 1962.

Though DNA composition of a species is fairly constant it occasionally undergoes mutation; the variants found in members of a population is principally due to this region. In organisms reproduction by asexual method mutations bring about variations. Mutations coupled with recombination *i.e.*, rearrangement of genes due to crossing over at the time of gametes formation results in variations found in sexually reproducing organisms.

EVIDENCES IN SUPPORT OF THAT DNA IS THE GENETIC MATTER

Even prior to the discovery of structure of DNA; it was well established that DNA is the genetic matter of living organisms. To prove that DNA is the genetic matter, experiments were carried out by the scientists during 1930s to 1940s.

Griffith's Experiment on Pneumococcus (Transformation)

Frederick Griffith a British doctor in 1928 performed some experiments with *Diplococcus pneunominae* commonly known as pneumococcus, a pathogenic bacterium that causes pneumonia in mammals including human beings. It was one of the first experiments suggesting that bacteria are capable of transferring genetic information through a process known as *transformation*. This bacterium like other living organisms has variants. There are two strains of *Diplococcus pneunominae*. One of the strains is known as *S-type*, which is virulent and causes disease pneumonia and the other strain known as *R-type* is avirulent and does not cause any disease. The other differences between these two strains of *Diplococcus* are given in table 10.1.

Table 10.1. Difference between S-type and R-type *Diplococcus pneunominae*

S.No.	*S-type*	*R-type*
1.	This strain of *Diplococcus* has a capsule composed of polysaccharides around the bacterial cell.	Capsule is absent around the bacterial cell.
2.	When this strain of the bacterium is grown in petri dishes on blood agar medium forms large smooth colonies.	This strain forms small rough colonies when grown on blood agar medium.
3.	This strain is pathogenic or virulent that causes death of mice when injected.	This strain is non pathogenic or avirulent and has no effect on mice when injected.

Griffith studied pathogenecity of pneumococcus on mice (Fig. 10.1) and made the following observations:

1. When live strain of S-type of *Diplococcus* was injected in mice; it developed pneumonia and the mice died after a few days.
2. Injection of the R-type of the bacterium in mice caused no ill effect on it and the mice remained alive.
3. Heat killed S type in which the bacterial cells were inactivated when injected in mice; it did not caused pneumonia and the mice remained alive.
4. When a mixture of heat killed S-type and live R-type was injected in mice; it caused pneumonia and resulted in death of the mice after a few days. Autopsies' of the mice revealed that death occurred due to the living bacterial cells of the virulent strain *i.e.*, S-type of *Diplococcus*.

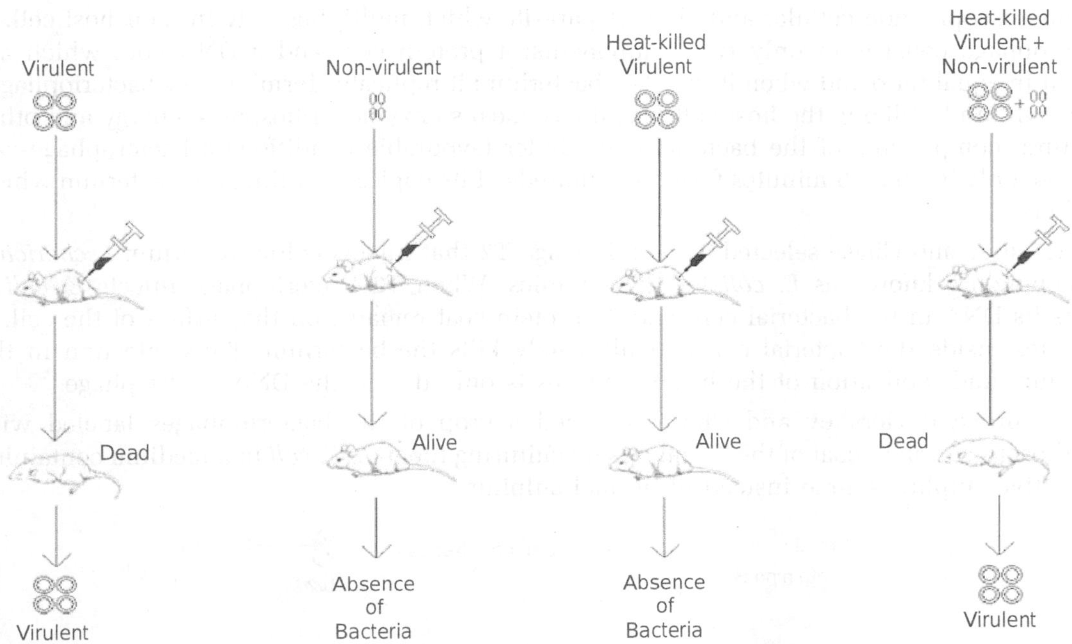

Fig. 10.1. Diagrammatic representation of Griffith's experiment on transformation in Diplococcus.

This fourth observation surprised Griffith and was unexpected; and he concluded that a chemical substance the *'transforming principle'* released by the heat killed and inactivated S-type of the bacterium transformed the R-type bacteria into S-type; and this was a permanent change because the transformed bacteria continued to produce similar bacteria.

Identification of the 'Transforming Principle'

The 'transforming principle' was identified as DNA in the year 1944 by **T Avery, MacLeod** and **McCarthy**. They first of all extracted and purified DNA, proteins and carbohydrates fractions from the S-type, virulent strain of *Diplococcus.* Then they cultured R-type of the bacteria mixed separately with the DNA, proteins and carbohydrates extracted from the S-type. It was observed that some of the R-type of bacteria cultured with the DNA fraction of S-type bacteria were transformed into S-type pathogenic bacteria but none of the R-type bacteria cultured with protein fraction or carbohydrate fraction of S-type bacteria was transformed into the S-type. This observation indicated that DNA is the *'transforming principle'* in the pneumococcus. When the R-type bacteria were cultured with the DNA of the S-type mixed with enzyme DNAase (the enzyme which degrades DNA, making it non functional) no transformation occurred in the R-type confirming that DNA is the genetic matter. During transformation an exogenous segment of DNA of a bacterium (donor) is incorporated in some other bacterium (recipient) to bring about genetic recombination in the recipient bacterium.

The Hershey-Chase Experiment

An additional and conclusive evidence to prove that DNA is the genetic matter was provided by **Alfred Hershey** and **Martha Chase** in 1952 with the experiments performed by them on bacteriophages. Bacteriophages are the type of viruses which kill bacteria. A virus is an ultramicroscopic, non cellular and obligate parasite which multiplies only in their host cell. A bacteriophage consists of only two components: a protein coat and a DNA core which are present in equal ratio and when it infects a bacterium; it replicates forming new bacteriophages of its own kind utilizing the host cell machinery means enzymes, ribosomes, energy and other structural components of the bacterial cell. Under favourable conditions a bacterophage can replicate only in 20 to25 minutes forming hundreds of its copies by killing the bacterium which it infects.

Hershey and Chase selected a bacteriophage T2 that infects colon bacterium *Escherichia coli* commonly known as *E. coli* for their studies. When T2 bacteriophage infects *E. coli* it injects its DNA in the bacterial cell; and its protein coat remains on the surface of the cell. It replicates inside the bacterial cell and ultimately kills the bacterium. Thus infection in the bacterium and replication of the bacteriophages is only due to the DNA of the phage.

To prove it Hershey and Chase obtained a crop of the bacteriophages labeled with S^{35}radioactive protein coat of the T2 phages by culturing them on *E. coli* in a medium containing radioactive sulphur source instead of normal sulphur.

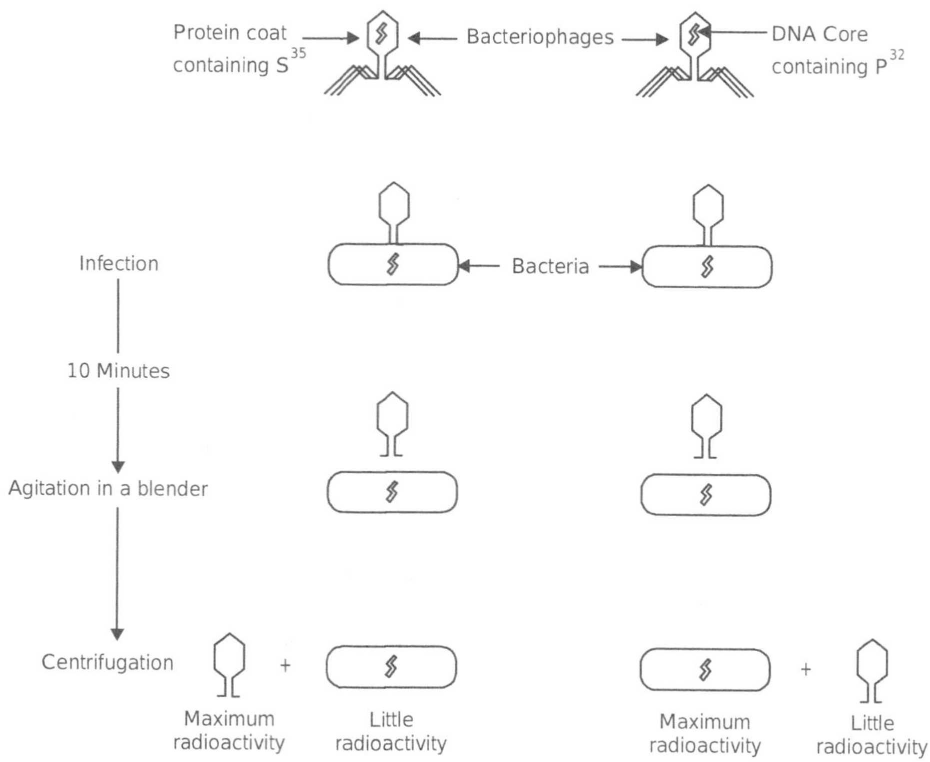

Fig. 10.2. Diagramatic representation of Hershey-Chase experiment to prove that DNA of bacteriophage is the genetic matter not its protein.

Similarly, they obtained P^{32} labeled DNA phages by culturing T2 phages on *E. coli* supplied with a radioactive source of phosphorous instead of normal one. This become possible as proteins contain sulphur but not phosphorous; and DNA contains phosphorous but not sulphur. When a T2 phage labeled with radioactive S^{35} protein coat was allowed to infect on an *E. coli* cultured on a normal medium, for a limited time (10 minutes) and after that they were subjected to a shearing force in a blender separating the protein coat and the bacteria. There after they were centrifuged and the proteins being lighter come in the supernatant on the surface and bacterial cells being heavier remained at the bottom of the centrifuge tube as a palate and they were separated. Their radioactivity was measured; it was found that radioactivity was associated with protein coat and almost none with the bacterial cell fraction. This indicated that no protein material of the phage had entered the bacterial cell in the course of infection process. The experiment was repeated by infecting the bacteria cultured on normal medium with P^{32} labeled DNA phages. Hershey and Chase found that in this experiment almost all the radioactivity was associated with the bacteria instead of protein coat fraction. These experiments proved that when a phage infect the bacterium it is only the phage DNA which enters into the bacterial cell and cause infection and protein coat has no role in it.The more important finding was that the bacteria which contained the injected DNA, but were separated from the phage protein coat, were capable of producing a fresh crop of normal bacteriophages. This experiment proved beyond doubt that the DNA of the bacteriophage is the genetic matter.

Evidence in Support of that RNA is the Genetic Matter in TMV

TMV is a rod shaped virus which infects tobacco and causes leaf mosaic disease in it. Besides causing disease TMV also replicates in its host forming many more virus particles or its progeny using the host cell machinery. The characteristic symptoms of this disease are the appearance of yellow patches on its leaves. TMV is composed of only RNA and proteins which

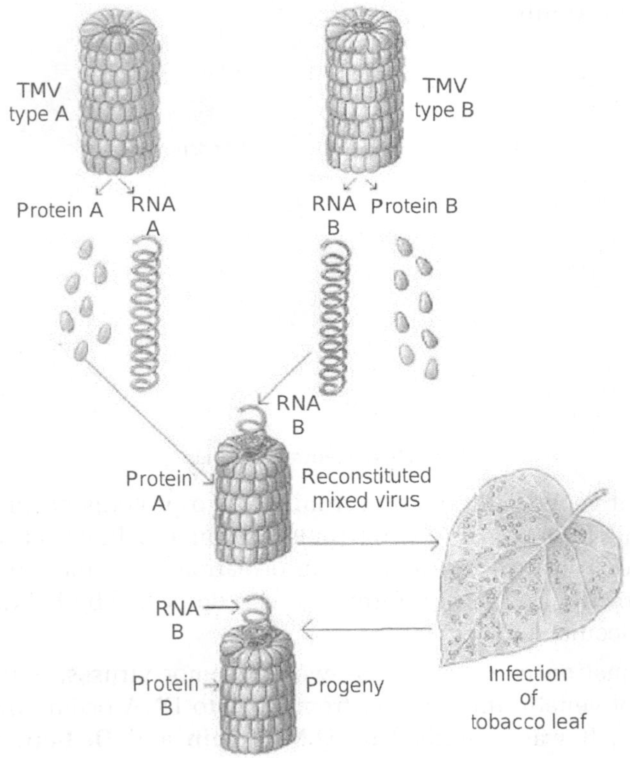

Fig. 10.3. Diagramatic representation of Fraenkel Conrat and Singer experiment showing that RNA of TMV is the genetic material

constitutes 6 percent and 94 percent of it respectively. RNA is present in the centre surrounded by protein capsomeres. It does not have DNA. The first experiment to prove that RNA of TMV is the genetic matter was conducted by **H. Fraenkel Conrat** and **B. Singer** in 1957. They separated RNA and proteins of TMV by some appropriate techniques and these were inoculated separately in the healthy tobacco plants. After a few days it was noticed that infection developed only in those plants which were inoculated with RNA not in the plants inoculated with proteins. This experiment showed that the infection is caused by RNA of TMV not by its proteins. Fraenkel Conrat and B.Singer also synthesized chimeric viruses using RNA from one strain and proteins from other strain yielding complete infective TMV particles. They took two different strains of TMV, separated RNA and proteins of each of the strains. Then RNA of one strain and protein of other strain were mixed; it resulted in the formation of 'mixed' virus or chimeric virus containing RNA of one strain and protein of the other strain. When these mixed viruses were used for infection on tobacco leaves, they caused the disease and also replicated forming progenies that had proteins corresponding to the parent strain from which RNA was obtained. So the genetic information of TMV is stored in its RNA not in its protein.

Central Dogma

The various experiments described above supporting that DNA is the genetic matter in all the living organisms except some viruses in which RNA is the genetic material instead has established that nucleic acids DNA or RNA is the store house of all the information which determines the characters of the living organisms. The information stored in the nucleic acids (DNA or RNA) is expressed through the synthesis of proteins.

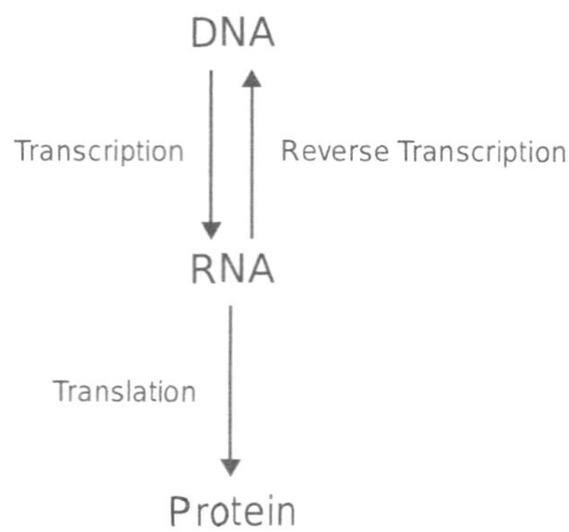

Fig. 10.4. Central dogma.

The expression of genetic information from DNA to proteins occurs in two consecutive steps: transcription and translation. During transcription the information stored in DNA is transferred to RNA which in turn transfers the information to proteins in the next step of the translation. This unidirectional flow of information was described by **F.H.C. Crick** (1958), as the **'central dogma'** of molecular biology.

The genetic information in retroviruses—wound tumor viruses, is transferred from RNA to DNA. The transfer of genetic information from RNA to DNA occurs by the process known as *reverse transcription*. It was described by **H.M. Temin** and **D. Baltimore** in 1970 and is carried out by RNA dependent DNA polymerase called reverse transcriptase. In the wound tumor viruses and other viruses like TMV, HIV (Human Immunodeficiency Virus) first DNA is synthesized on the RNA and then this DNA in turn synthesize RNA(transcription) and RNA in turn synthesizes proteins (translation).

Genetic Engineering

Genetic engineering means *manipulation of genes at will.* Manipulation of genes is achieved either by addition, removal or repair of a part of genetic material by artificial means there by changing the phenotype as desired. Now it is possible to isolate any gene of interest from any species and it can be incorporated into any another species by forming a recombinant DNA by uniting the isolated gene of interest either with a plasmid (extra chromosomal DNA of bacteria) or viral DNA and then transferring it into a host cell. When a recombinant DNA is introduced in a host cell it propagates and multiplies as the cell divides to form new cells. This technology enables the gene transfer between totally unrelated species. Genetic engineering bypasses the restriction in the gene transfer mechanism between unrelated organisms. By this technology it is possible to transfer a gene from human cell into a bacterium in which this gene expresses itself and produces the same type of products which it produces in humans.

RECOMBINANT DNA TECHNOLOGY

Recombinant DNA is a type of synthetic DNA that is formed outside living cells by joining DNA segments from two or more sources that can replicate in living cells (Fig. 10.5). Recombinant DNA technology was developed by **Stanley Norman Cohen** and **Herbert Boyer** in 1973. They described the process of isolation and amplification of genes or DNA segments and insert them in another cell. They generated transgenic bacteria by this technique. The recombinant DNA technology was made possible by the discovery of restriction enzymes (endonucleases) by **Werner Arber, Daniel Nathans**, and **Hamilton Smith,** for which they received the 1978 Noble Prize in Medicine. The technique to obtain a recombinant DNA is as follows:

(*i*) Extraction of gene of interest from DNA of donor organism taken from its cells. The gene of interest from donor's DNA is separated with the help of restricting enzymes (endonucleases) known as biological scissors. The restriction enzymes cut the DNA at specific places so that a particular gene can be isolated by appropriate methods.

(*ii*) This isolated gene can be introduced into a cell of another organism using a vector. A vector is a 'vehicle 'by which gene is transferred from one cell into another.

(*iii*) A plasmid can be used as a vector. Plasmid is an extra chromosomal and circular DNA present in certain bacterial cell.

(*iv*) The plasmid is also treated with the same type of restriction enzyme that is used to separate the gene of interest from the donor organism so that gap in the circular DNA opens to combine with the new pieces of DNA (gene).

(*v*) The restriction enzyme cuts the same site on both molecules. The ends of the cut have an overhanging piece of single stranded DNA. These are called *'sticky ends'*.

(*vi*) Now the gene is combined with the plasmid. This becomes possible as the 'sticky ends' are able to base pair with any DNA molecule containing the complementary 'sticky ends'. An enzyme known as ligase helps in linking the two into a molecule of recombinant DNA.

(*vii*) The recombinant piece of DNA now contains the new gene which can be introduced in another cell.

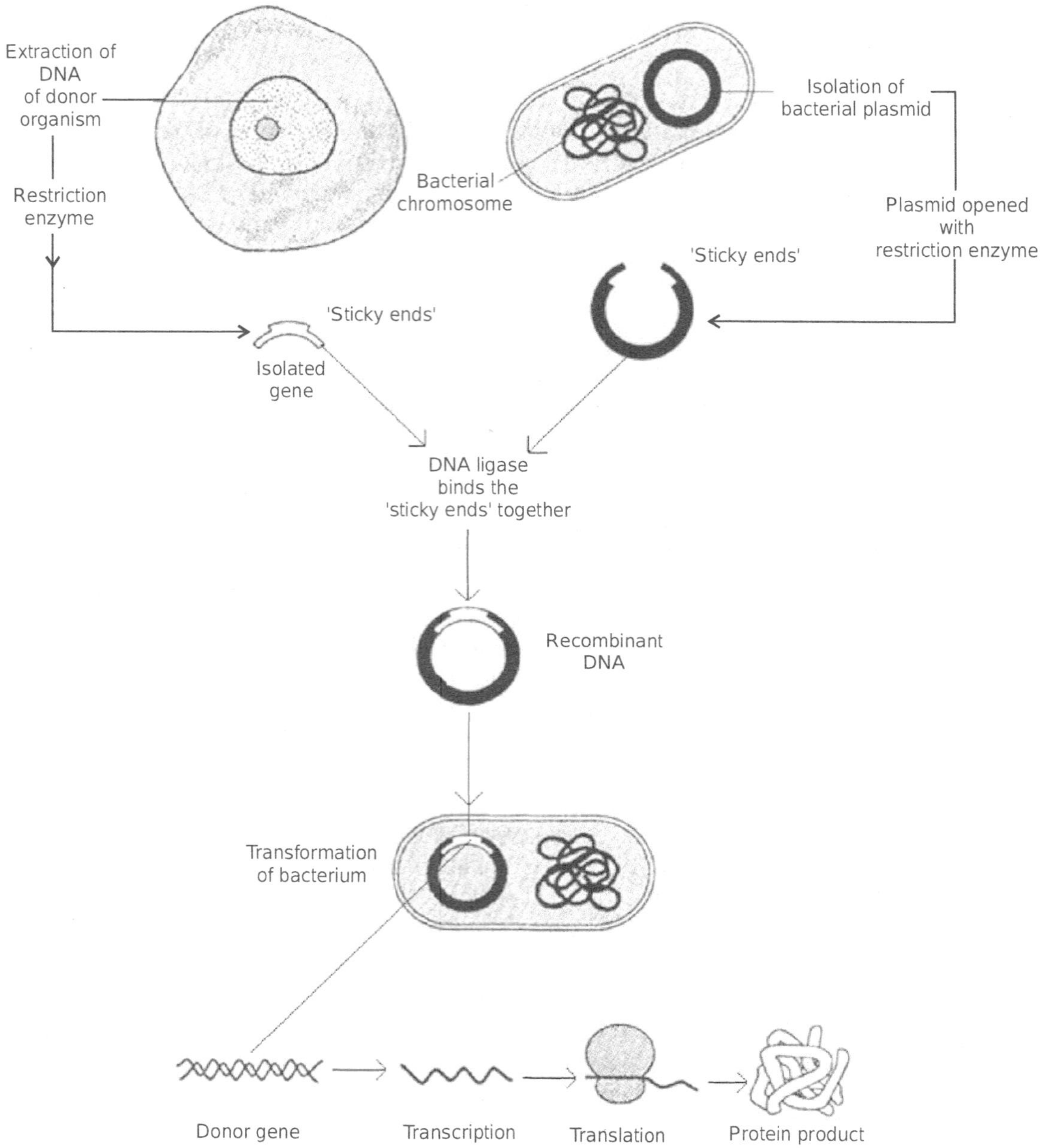

Fig. 10.5. Diagrammatic representation of recombinant DNA technology.

Application of Recombinant DNA Technology

Recombinant DNA technology is of great biological significance. It has many practical applications at present and has many prospects in future.

(*a*) It is utilized to obtain many important metabolites like hormones, antibodies etc. Large scale production of hormones like insulin is possible by introducing genes from higher animals coding for it into bacteria; as bacteria multiply very rapidly thus creating 'living factories' from which insulin can be extracted. At present insulin is obtained from pancreas of animals like mice which has so many limitations.

(*b*) In future recombinant DNA technology may be used to cure hereditary diseases like haemophilia, phenylketonuria etc. by correcting defective genes by gene therapy which is otherwise incurable.

(*c*) There is a possibility of transferring of nitrogen fixing genes from bacteria or blue green algae into cereals like wheat, rice, maize etc. This will help in increasing the production of these crops and decreasing the cost of production. As a result of which synthetic fertilizers will not be required and so will also help in checking the water and soil pollution.

(*d*) Transgenic plants are obtained by this technique which are pest resistant thus increasing their yield.

(*e*) Transgenic fishes with high protein contents are obtained by recombinant DNA technology.

(*f*) Transgenic pigs have been developed containing human genes. The organs of transgenic pigs can be used to replace defective body parts. This method of transplantation is at the threshold of beginning of a new era in the field of medicine.

GENE LIBRARY

Gene library means a collection of cloned DNA fragments from a variety of organisms. To set a gene library of an organism its DNA is isolated from its cells or tissue. The isolated DNA is cut into small fragments with the help of restriction enzymes. Each of the fragments contains one or a few genes. These fragments are integrated separately into isolated bacterial plasmids. Plasmid is a small circular DNA and it is also opened by treating with the restriction enzymes. To join an opened plasmid with a fragment of DNA of the organism is done by the help of the enzyme ligase to form recombinant DNA molecules. These recombinant DNA molecules are introduced in suitable host cells like a bacterial, plant, or animal cells to generate transgenic cells. Each of the transgenic cells has one or a few genes of the organism whose gene library is to be set. These cells are cultured in a suitable medium so as to generate clones (exact or carbon copy) of DNA fragments of the organism. In the similar fashion gene library of any organism can be set and there are two types of gene libraries: Genomic library and c-DNA library.

(*i*) **Genomic Library.** A genomic library is a collection of clones representing the complete genome of an organism. This means that entire set of genome of an organism is represented as a set of DNA fragments inserted or cloned into a vector molecule that can be propagated in a suitable host. It is set as explained in the general process of setting of a gene library.

(*ii*) **c-DNA Library.** In case of a eukaryotic organism gene library is not set with its DNA but is made with the help of mRNA isolated from its cells. As DNA of eukaryotic organisms is consists of introns and exons. Introns are the non transcriptional part of the DNA. So gene library in eukaryotes is made with the help of mRNA. In eukaryotes, mRNA is formed by splicing *i.e.,* removing those parts of Hn-RNA (Hn=heterogeneous) corresponding to intron regions of DNA. With the help of isolated mRNA a single stranded c-DNA is generated with the help of the enzyme reverse transcriptase and it is then duplicated by appropriate technique. Once c-DNA molecules are generated, they are cloned in suitable host cells by integrating them in a vector.

GENOMICS

Genomics is the study of organism's entire genomes. A genome is the sum total of all the genes which an organism possesses. The genome of every organism is the repository of all its genetic information and it is basically stored in the individuals DNA. Genomics is essentially the study of genes and their function in relation to environment. The term genomics was coined by **Thomas Roderick** in 1986. The various activities undertaken in genomics include mapping, sequencing and analyzing genomes. Genomics was practically founded by **Fred Sanger** and coworkers in 1970s when they develop a gene sequencing technique and first of all completed the sequencing of the genome of bacteriophage ϕX174, human mitochondrial genome and genome of lambda virus. Thus the field of genomics includes intensive efforts to determine the entire DNA sequences of organisms and making genetic maps of all of their genes. The goal of genomics is to promote the understanding of the structure, function, and evolution of genomes of all kingdoms of life and the application of genome science and technology to the challenging problems in biology and medicine. Till now genomes of about 60 different organisms including human being have been completed.

An ambitious Human Genome Project began in 1990 that was coordinated by the US in collaboration with other countries like England, Japan, France, Germany, China and others have been completed. The important goals of this project were.

1. Identification of total genes present in human DNA and mapping of the human genome.
2. To store this information in databases and develop tools for its analysis.
3. To address the ethical, legal and social issues that may arise from this project.

Though human genome project has been completed analysis of its data will continue for many years. Some of its findings have revealed startling.

1. Human possesses only about 30,000 genes. In the beginning of this project it was expected that humans may have about 100,000 genes. We have just twice the number of genes as that of *Drosophila*, an insect; six times more than *E. coli,* a bacterium; and almost the same number of genes as that of mouse and moreover nine-tenths of our genes are identical as that of mouse.
2. It has also revealed that 99.9 percent DNA of all the human beings is identical and it is only .1 percent DNA which different and this minor difference is the cause of dissimilarities in the features and other characters of human beings.
3. It has also revealed that out of the total human DNA only less than 2 percent of it is known to include the exons, the protein coding sequences the rest constitutes introns or the non coding sequences. This type of DNA is present as base pair repeated sequences clustered around, the centromere and near the ends of the chromosomes and is referred as *'the junk DNA'*.

Human genome project has enabled to gather a wealth of information that will ultimately play an important role in preventing, diagnosing and managing a variety of chronic diseases including cancer, diabetes, and cardiovascular diseases.

Bioinformatics

Bioinformatics is the application of computer technology to the management of biological information. Computers are used to gather, store, analyze and integrate biological and genetic information. Bioinformatics comprises the use of techniques from applied mathematics, informatics, statistics, and computer science to solve biological problems. The primary aim of it is to increase our understanding of the biological processes. The ultimate goal of the field is to enable discovery of new biological insights as well as to create a global perspective from which unifying principles in biology can discerned. The term *bioinformatics* was coined by **Paulien Hogeweg** in 1978 for the study of informatics processes in the biological systems. Bioinformatics encompasses the creation and maintenance of database to store biological information. Development of this type of database not only design issues but leads to development of complex inferences where by researchers could both access existing data as well submit new or revised data. Major research areas in bioinformatics include sequence alignment, gene finding, genome assembly, protein structure prediction, and prediction of gene expression, protein-protein interaction and the modeling of evolution.

Exercises

1. What is the genetic matter of living organism?
2. Describe the Griffith's experiment on transformation.
3. Name the scientists who proved first of all that DNA is the genetic matter.
4. Describe the Hershey and Chase experiment to prove that DNA is the genetic matter of bacteriophage.
5. What is the evidence in support of RNA as the genetic matter of viruses.
6. Write a brief note on central dogma.
7. What is recombinant DNA technology?
8. Write a note on genetic engineering.
9. What is a gene library and what is its importance
10. What is genomics? What are the prospects of genomics?
11. Write a note on human genome project.
12. Define bioinformatics and what is its significance.

❑❑❑

UNIT 4

- ★ Biotechnology: Scope, Achievements and Techniques
- ★ Applications of Biotechnology

Chapter 11 Biotechnology: Scope, Achievements and Techniques

Biotechnology is one of the most exciting and revolutionary science of this century. However, biotechnology is not a new branch of science. Biotechnology is as old as human civilization. It has shown its presence in one form or other and is in practice since long to generate goods of utility by encompassing biological processes and the technology. It has flourished in one form or the other since the advent of different civilizations of the world. Biotechnology broadly speaking is the use of living organism particularly microbes or their components to generate goods and services useful to human beings. Biotechnology is a term coined by the combination of two words: *Biology* and *Technology. Biology* is the study of living organisms, and *technology* means any technique or process by which raw materials are converted into products; and when living organisms are involved in the conversion of raw materials into products it is known as **biotechnology.** The term biotechnology dates from 1919, when the Hungarian engineer **Karl Ereky** used it to mean *'any product produced from raw materials with the aid of living organisms'*. But humans are using microorganisms for the production of bread, curd, yogurt and alcoholic beverages from the prehistoric time and there are references of their use in different civilizations. Ancient civilizations were using the microorganisms to make useful products, although of course, they had no idea that it was microbes that were the active agents. Long back Sumerians and Babylonians discovered how to use yeast to make beer and wine. Thousands of years back the Egyptian found that addition of yeast produced fluffy bread instead of thin, hard wafer. At the same time the Chinese were adding bacteria to produce yogurt. Use of curd and 'makhan' (raw butter) and 'lassi' (milk water) is in practice in India since time immemorial and the lactic acid bacteria brings about conversion of milk into curd and from the curd related food items as mentioned above are obtained. Similarly, the use of fermented foods like 'idli' and 'dosa' are very common in southern and the use of 'jalebi' and 'imrti' common in northern parts of our country, which requires the application of microbes at some steps of their preparation.

Microorganisms are employed not only in the processing of food and formation of alcoholic beverages as mentioned above but are also used in some other processes like extraction of fibers from the fiber crops like jute and sun hemp by retting, and curing of tea and tobacco, tanning of leather etc. Many chemical compounds used in varied fields are obtained from microorganisms. **Weizmann** in 1917 found that corn starch can be converted into acetone with the help of a bacterium, *Clostridium acetobutylicum*. Acetone was used to manufacture explosives during World War-I by the United Kingdom. Antibiotics were discovered by **Alexander Fleming** in 1929. He isolated the first antibiotic penicillin from fungus *Penicillium notatum*. Later **Waksman** isolated streptomycin from a bacterium *Streptomyces*. The antibiotics, first time come in prominence during World War-II when thousands of wounded soldiers were saved from the infection by the treatment of antibiotics.

Genetic engineering or recombinant DNA technology discovered in 1970s along with the tissue culture technique which get fully established in 1950s has given varied dimensions and different meanings to biotechnology.

DEFINITION OF BIOTECHNOLOGY

At present a single definition is not sufficient to explain its meaning but many of its definitions are prevalent as suggested by different scientists and organizations from time to time. Some of these definitions are:

According to the UN-Convention of Biological Diversity, biotechnology may be defined as, *"any technological application that uses biological systems, living organisms, or derivatives of thereof, to make or modify products or processes for different uses"*.

According to The Organization for Economic Cooperation and Development (OECD), *"biotechnology is the application of scientific and engineering principles to the processing of materials by biological agents to provide goods and services"*.

According to the European Federation of Biotechnology (EFB), biotechnology is defined as, *"the integrated use of biochemistry, microbiology and engineering sciences in order to achieve technological application of the capabilities of microorganisms, cultured tissues, cells, and parts their of"*.

According to Japanese biotechnologists it is defined as, *"a technology using biological phenomena for copying and manufacturing various kinds of useful substances"*.

SCOPE AND ACHIEVEMENTS OF BIOTECHNOLOGY

Recent rapid advances in genetic engineering, cell culture and molecular biology have generated a virtually unlimited potential for altering the capabilities of living systems, greatly expanding the scope of biotechnology, generating new applications for biological products and allowing an unprecedented ability to control life processes. Biotechnology is of great significance that has proved its worth in almost every field concerning to human welfare. It has already contributed a lot and has great scope in future in the field of agriculture, industry, health care and management of environment.

Biotechnology has lots of scope in agriculture and animal husbandry. Although green revolution that started in 1960s has contributed a lot to increase the food production by way of producing high yielding crop varieties. But the use of agricultural inputs like chemical fertilizers, pesticides, and fungicides has increased the cost of production and at the same time has put a heavy pressure on the fast depleting natural resources like the fossil fuels; petroleum and natural gases that are required for the production of chemical fertilizers, pesticides and fungicides. Besides, it the large scale use of these agricultural chemicals has resulted in polluting the environment (soil, water and air).

Biotechnology has to contribute a lot in ameliorating such problems as it encourages organic farming. Employing biotechnological techniques in agriculture has already shown many fruitful results like increase in quality and quantity of crop production, reduction in the cost of agricultural inputs and decrease in the dependence on chemical fertilizers and use of fungicides, pesticides etc. At present a number of biopesticides, biofertilizers, and biofungicides have been developed which are environment friendly and in their production no fossil fuel is required. But at present the use of these products is limited as they are slow in action but are long lasting. Hopefully, the acceptability of use of these bioproducts by the farmers would get momentum in the times to come once they learn its importance.

In the recent times many transgenic crop plants have been developed. Important transgenic crop plants include Bt corn and Bt cotton (pest resistant), Flavr Savr tomato (increased shelf life), Golden rice (rich in vitamin-A), herbicide resistant soybean and viral disease resistant tobacco. Efforts are on to introduce a nitrogen fixing gene from bacteria like *Rhizobium* and *Azospirillum* into cereals like wheat and rice. Similarly vigorous research is going to generate transgenic salt tolerant pulses like chick pea and green gram. At present due to some reasons or the other the use of transgenic crop plants is limited. There is a lot of scope of their popularization and large scale use in future that will definitely start a new era of farming, the organic farming in which use of chemicals will be limited.

As far industries are concerned a powerful industrial base has been developed in recent times based on culturing of microorganisms. A careful selection of microbial strains coupled with improved methods of extraction and purification have resulted in enormous yield. Brewing and baking industries which are in existence since the advent of the human civilization are based on the activity of a unicellular fungus, the yeast. Brewing industries are related to the production of alcoholic beverages like beer, wine, brandy and scotch. Baking industries are involved in the production of bread, cakes, pastries etc. Brewing and baking industries have achieved a spectacular growth in the recent times by refining the process and use of the improved strains of yeast. Many an industrial products like acetone, gluconic acid, citric acid, vinegar, enzymes, vitamins etc. which are the raw materials needed for various purposes are obtained by culturing of microbes and biotechnological techniques.

Biotechnology is of great significance in the field of medicine and health care. Recombinant DNA and biotechnology have opened a new era of diagnostic testing and have made detection of many genetic diseases possible which were difficult to detect in the past. Many a products of recombinant DNA technology are in use to treat that disease which was difficult to treat or incurable in the past. For example, insulin is used for the treatment of diabetics, human factor VII for the treatment of haemophilia-A, human factor IX for the treatment of haemophilia-B, and hepatitis B vaccine for prevention of hepatitis-B.

To reduce pollution and managing the other environmental problems biotechnological techniques which uses microbes is contributing significantly and also has a great scope in future.

TOOLS AND TECHNIQUES OF BIOTECHNOLOGY

Biotechnology basically employs the knowledge gained in the field of molecular biology, genetic engineering, microbiology, biochemistry, cell culture etc. with the help of which a number of techniques have been evolved and these techniques are used to obtain products of human welfare. Main activity of modern biotechnology is based on recombinant DNA (rDNA). The various techniques used in biotechnology are as follows:

Plant Tissue Culture and Biotechnology

Plant tissue culture is the technique of culturing *i.e.,* growing of isolated cells, tissues or organs of plants on nutrient medium under aseptic and controlled environmental conditions in the laboratory to raise new plants. Plant tissue culture technique is based on the concept of cellular totipotency. Totipotency means ability of a cell to form a complete plant from it. This concept of totipotency was first conceived by **Haberlandt** (1902), a German botanist, who cultured isolated palisade cells of leaves in Knop's solution enriched with sucrose. The cells remain alive up to one month increased in size but failed to divide. However, this attempt to demonstrate cellular totipotency, ultimately leads to the development of techniques for culturing plant cells in laboratory under aseptic controlled environmental conditions. Development of somatic embryos was first reported in 1958–1959 from carrot tissues independently by **Reinert** and **Steward.**

Technique of Plant Tissue Culture

Plant tissue culture technique enables the growing of explants *i.e.,* isolated cells, tissue or organs taken from a desired mature plant. Explants are grown on nutrient media under aseptic and controlled environmental conditions to obtain new plants. The salient points of tissue culture technique are summarized here and the requirements and procedure followed is given below:

(a) Requirements

Essential requirements for tissue culture technique are: explants, nutrient medium, glassware (test tube, flasks, petri dishes etc.), oven, autoclave, laminar flow chamber, metal instruments(scalpel, scissors, inoculation needle), and chemicals (alcohol, mercuric chloride, sodium hypochlorite) and a culture room.

(*i*) *Explants.* Explants are obtained from a desired mature plant. It may be shoot tip, root tip, cambial cells etc. which are cut by a scissors or a scalpel and are separated from the parent plant and are collected in a suitable container like a test tube or Petri dish.

(*ii*) *Nutrient Medium.* A nutrient medium is consists of various essential macro and micronutrients, a carbon source in the form of sugars, vitamins like biotin, niacin and growth hormones auxins, cytokinins and gibberellins. Solidifying agent agar-agar (a gelatinous polysaccharide obtained from *Geladium*, an alga) is generally added into the medium to make it semisolid. There are various types of culture media the most commonly is culture medium is MS (Murashige and Skoog) medium. Sometimes coconut milk and corn extract is also added in the medium. The medium is filled in suitable containers like test tubes, flasks and are properly plugged. The test tubes and flasks containing culture medium are properly sterilized either in an autoclave or in a pressure cooker.

(*iii*) *Oven.* It is used to sterilize glassware and metal instruments. Required glassware and metal instruments are kept in an oven for about one hour and the temperature is set at 200^0C to sterilize them.

(iv) *Autoclave.* An autoclave is used to sterilize the nutrient medium, distilled water etc. Sterilization is done at a temperature120°C and 15 lb pressure for about 30 minutes so that the nutrient medium and distilled water becomes totally free of microbes.

(v) *Laminar Flow Chamber.* Inoculation of explants and sub culturing is done in a laminar flow chamber in which there is flow of sterile air, free of microbes. Working in such chambers for culturing of plant tissue ensures a microbe free environment and to check contamination.

(vi) *Culture Room.* Culture room maintains proper temperature, humidity and light which are required for the purpose of tissue culture and growth of plants obtain by tissue culture technique. Proper care is taken so as to avoid any contamination.

(b) Procedure

(i) *Surface Sterilization of Explants.* The selected explants (plant cells, tissues or organs) which are to be cultured are first of all surface sterilized. Surface sterilization is done by treating them with a disinfectant like 0.5 percent mercuric chloride or 1.0 percent sodium

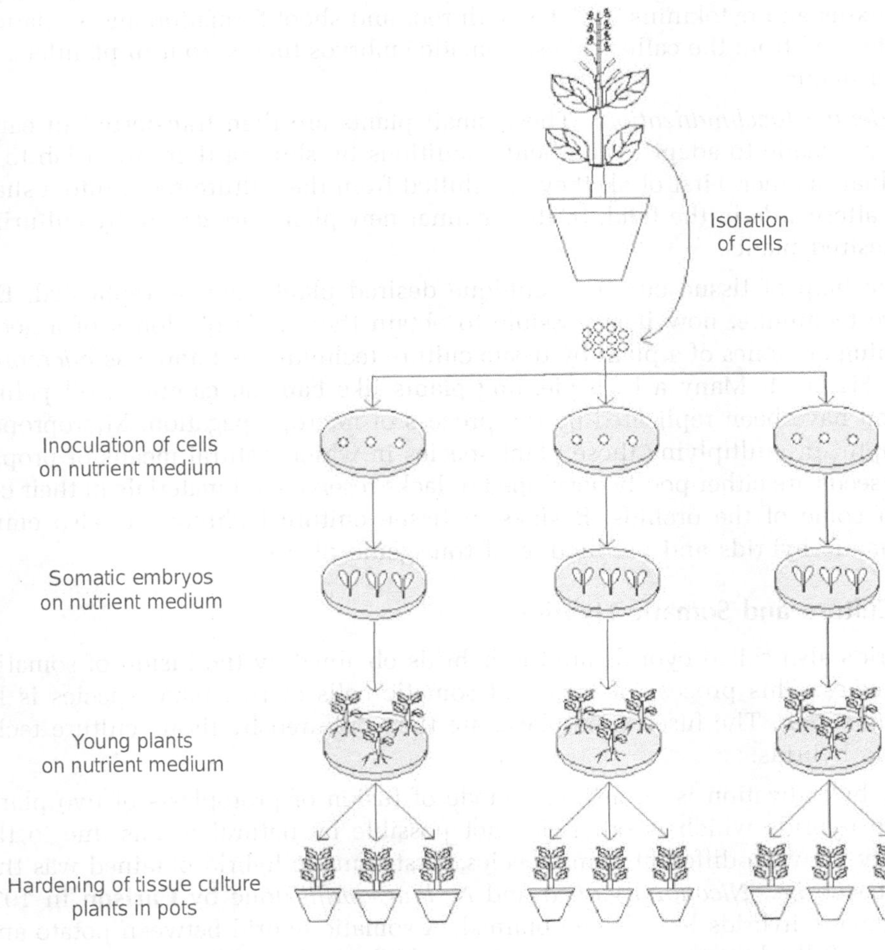

Fig. 11.1. Tissue culture technique (micropropagation).

hypochlorite solution. Then the explants are properly washed with sterile distilled water. It ensures checking of the contamination of microbes and making the explants microbes free.

(ii) *Inoculation.* Properly sterilized explants are transferred on the culture medium. This is known as inoculation. For this purpose culture medium is filled in suitable containers like petri dishes, test tubes or flasks and it is sterilized then the explants are put on the medium with the help of a sterile forceps or an inoculation needle this is known as inoculation. This operation is done in a laminar flow chamber under aseptic conditions.

(iii) *Incubation.* The petri dishes, test tubes or flasks containing the explants (excised plant tissues) are incubated *i.e.,* allowed to grow in an environment where a proper arrangement of light (16 hour photoperiod), temperature (about 26°C) and humidity (50–60 percent) is maintained in a culture room. Generally callus, an unorganized mass of cells, is formed from the explants. Organogenesis is initiated from the callus by treating it with growth hormones. Auxins application to callus initiates root formation whereas treating it with cytokinins initiates shoot formation. A balanced application of auxins and cytokinins initiates both root and shoot formation and somatic embryos are formed from the callus. These somatic embryos then sprout in plantlets, *i.e.,* small sized plants.

(iv) *Hardening (acclimatization).* These small plants are then transferred in earthen pots and are made to adapt to the field conditions by shifting them from lab to land in a gradual manner. First of all they are shifted from the culture room into a shady place, and afterwards in the field. In this manner new plants are grown by culturing tissues of desired plants.

With the help of tissue culture technique desired plants can be replicated. Employing tissue culture technique, now it is possible to obtain thousands of clones of a good quality plant. Obtaining of clones of a plant by tissue culture technique is known as *micropropagation* as shown in Fig. 11.1. Many a high yielding plants like banana, camphor, oil palm, mango, papaya, lemon have been replicated by the process of micropropagation. Micropropagation is also very helpful in multiplying those plant species in which natural means of propagation is slow as their seeds are either poorly developed or lacks reserve food materials in their cotyledons as in case of some of the orchids. Besides, it tissue culture technique is also employed in obtaining somatic hybrids and production of transgenic plants.

Protoplast Culture and Somatic Hybrids

Somatic hybrids also called cybrids are the hybrids obtained by the fusion of somatic cells of two plant species. This process of fusion of somatic cells of two plant species is known as *somatic hybridization.* The fused protoplasts are then cultured by tissue culture technique to obtain somatic hybrids.

Somatic hybridization is an artificial mode of fusion of protoplasts of two plant species to obtain rare hybrids which is otherwise not possible by natural means due to the sexual incompatibility between different plant species. First somatic hybrid obtained was that of two species of tobacco *i.e., Nicotiana glauca* and *N. langsdorffii* done by **Carlson** in 1972. Since then many somatic hybrids have been obtained. A somatic hybrid between potato and tomato has been successfully done to generate pomatoes, which combines the characters of both potato and tomato. Technique to obtain somatic hybrids is as follows:

1. First of all cells of the two plants whose somatic hybrid is to be formed are isolated from the respective plants.
2. Cells of both of the plants are treated with enzymes like pectinase and cellulase to get their protoplasts. Protoplast is a plant cell without cell wall.

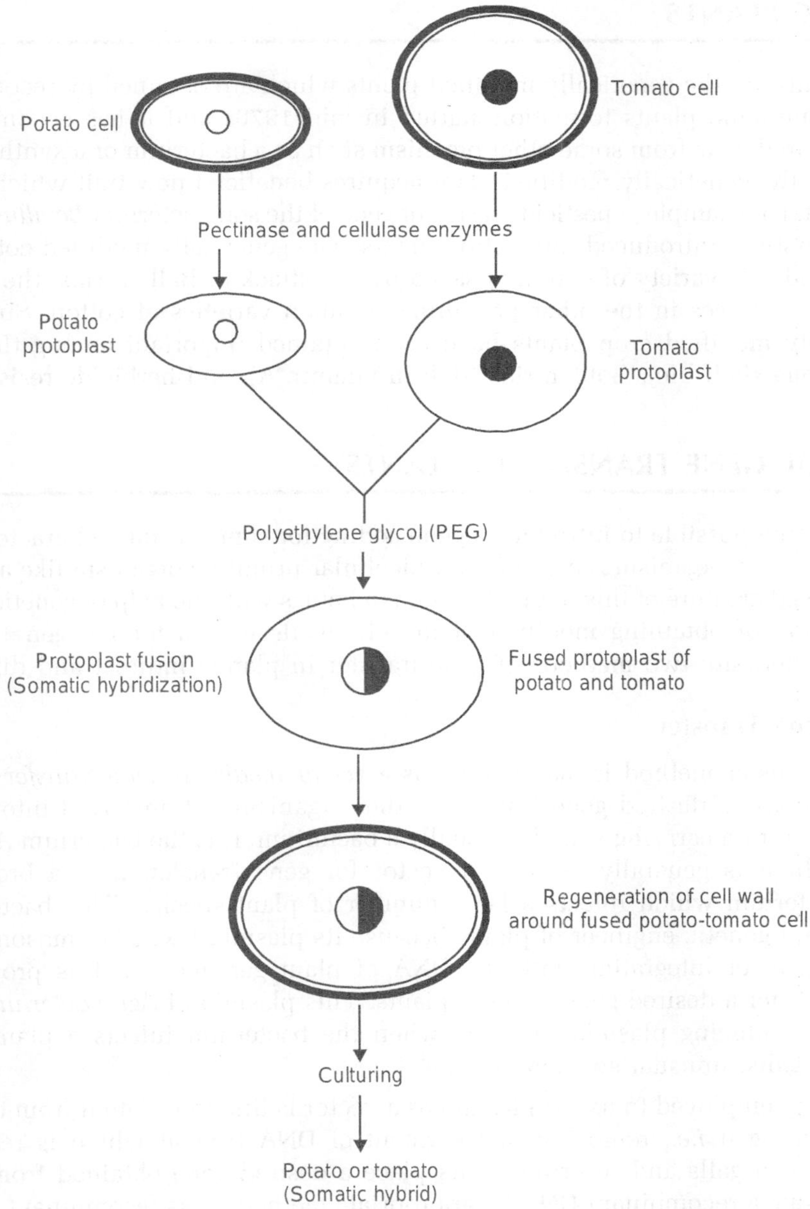

Fig. 11.2. Technique of protoplast fusion to obtain somatic hybrids.

3. The protoplasts of both the plant cells are put together in a suitable container. They are fused either by treating them with PEG (polyethylene glycol) or subjecting them to mild electric shocks.

4. The fused protoplasts regenerate cell wall and are shifted on a nutrient medium. They are cultured under aseptic and controlled environmental conditions as described in the technique of plant tissue culture.

TRANSGENIC PLANTS

Transgenic plants are the genetically modified plants which are obtained by recombinant DNA technology. Transgenic plants formation started in mid 1970s and it has become possible to introduce a desired gene from some other organism such as a bacterium or a synthetic gene into a plant, so that the genetically modified plant acquires beneficial new trait which is otherwise not present in it. For example, a pesticide resistant gene of the soil bacterium *Bacillus thuringiensis* has been successfully introduced into cotton plants. The genetically modified cotton is known as Bt Cotton and this variety of cotton is saved by the attack of Ball worms, the insect which causes large scale losses in the other prevailing common varieties of cotton. Similarly, many other genetically modified crop plants have been obtained important among them are Flavr Savr tomato (long shelf life), Golden rice (rich in vitamin-A), and herbicide resistant soybean.

METHODS OF GENE TRANSFER IN PLANTS

Now it has become possible to introduce a gene coding for some specific character into a plant from any other living organism ranging from unicellular primitive organism like a bacterium to the highly evolved creature of this world, the human beings with the help of genetic engineering. The phenomenon of obtaining modified plant cells (cells with a foreign gene) is known as transfect ion. There are two methods of gene transfer in plants: indirect and direct.

(*i*) Indirect Gene Transfer

Indirect gene transfer method is also known as a *vector mediated gene transfer* method (Fig. 11.3) as in this case a desired gene from any other organism is transferred into a plant with the help of a vector (a carrying vehicle); usually a bacterium. It is the bacterium *Agrobacterium tumefaciens* which is generally used as a vector for gene transfer. It is a broad spectrum pathogenic bacterium which infects a large number of plant species. This bacterium is also known as natural genetic engineer of plants because its plasmid (extrachromosomal DNA) has a natural property of integrating with the DNA of plant genome and this property of it is exploited to transfer a desired gene into the plants. This plasmid of *Agrobacterium* is called Ti plasmid (tumor inducing plasmid) because when the bacterium infects a plant its plasmid induces crown galls; unusual swollen areas.

The strategy employed to use Ti plasmid as a vector is first to isolate it from the bacterium and then disarming it *i.e.*, removing that segment of DNA from it which is responsible for formation of crown galls and inserting in its place a desired gene obtained from some other species to generate a recombinant DNA by appropriate method. This recombinant DNA is again put in the bacterium. Such modified bacteria are allowed to infect plant cells. For this purpose usually leaf discs of the plant are taken and are infected by the *Agrobacterium*. On infection the Ti plasmid transfer the foreign gene into the plant genome and these cells are cultured by the usual plant tissue culture technique as described in a previous section. After a month or so genetically modified plants emerges out which are carefully acclimatized. These plants are then grown in the field.

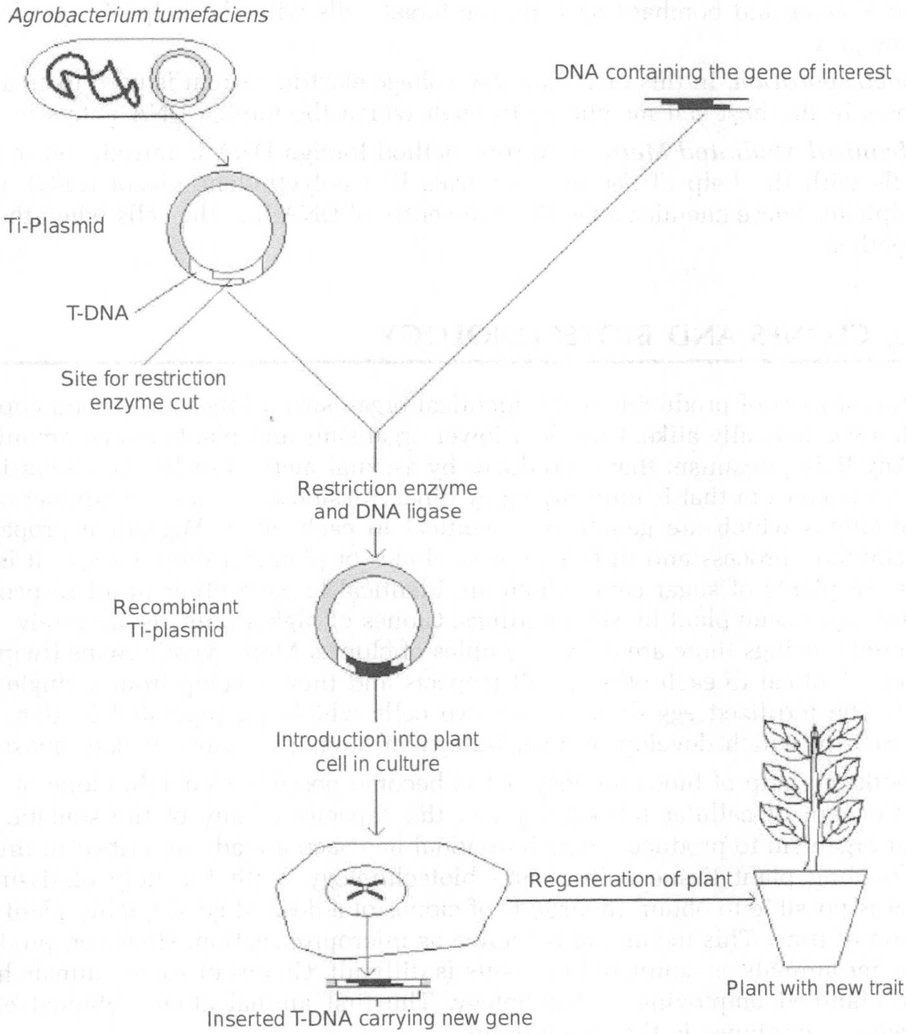

Fig. 11.3. Technique of obtaining transgenic plants employing *Agrobacterium*.

(ii) Direct Gene Transfer

Direct gene transfer is done without the use of a vector. In this method a DNA fragment containing gene or genes of interest which are either isolated from some other organism by appropriate technique or are artificially synthesized are transferred into plant cells. The plant cells which acquire the foreign genes are selected by suitable markers and are cultured by the usual plant tissue culture technique to obtain the transgenic plants. There are a number of direct gene transfer methods like microinjection, particle gun, electroporation and chemical mediated gene transfer method.

(*a*) ***Microinjection.*** In this method exogenous DNA is introduced directly into the plant cell with the help of a fine tipped 0.5–0.1 micrometer diameter glass needle or a micropipette.

(*b*) ***Particle Gun.*** In this method foreign DNA containing the gene of interest is introduced in a plant cell by coating it on minute gold or tungsten particles of the size 1–3 micrometer and bombarding it on the target cells with the help of a particle gun or gene gun.

(*c*) ***Electroporation.*** In this method a low voltage electric current is used to create minute pores in the host cell membrane through which the foreign DNA enters in the cells.

(*d*) ***Chemical Mediated Method.*** In this method foreign DNA is introduced in the target cells with the help of certain chemicals like polyethylene glycol (PEG) or dextran sulphate. These chemicals facilitate the entry of DNA into the cells when they are put together.

CLONING, CLONES AND BIOTECHNOLOGY

Cloning is the process of producing many identical organisms. Clones are carbon copy of each other which are genetically alike. Clones of lower organisms and plants are commonly formed in nature. Any living organism that reproduces by asexual method in fact produces its clones. For example, a bacterium that is multiplying by binary fission forms a large number of bacteria that are the clones which are genetically identical to each other. Vegetative propagation in plants is a common process and in this process clones of plants are formed *e.g.,* it is possible to obtain 10–15 plants of sugar cane which are identical to each other in all respects from a single mature sugar cane plant by stem cuttings. Clones of higher animals are rarely formed in nature. In human beings there are a few examples of clones. Monozygotic twins (twin brothers or sisters) are identical to each other in all respects and they develop from a single fertilized egg (zygote). The fertilized egg divides into two cells which get separated in their mother's womb and each of which develop into separate twin having the same genetic make up.

Now with the help of biotechnology it has become possible to obtain clone of any plant or animal. Concept of cellular totipotency *i.e.,* the capacity of any of the somatic cell of a multicellular organism to produce a new individual has been already described in this chapter under the heading plant tissue culture and biotechnology. With the help of tissue culture techniques it is possible to obtain thousands of clones of a desired good quality plant in a very short duration of time. This technique is known as micropropagation. However, production of clones of higher animals as compared to plants is difficult. Clones of some animals have been successfully obtained employing biotechnology. The first animal clone obtained employing biotechnological techniques is the sheep dolly.

Dolly (sheep)

Dolly, the first mammal to be cloned (Fig. 11.4) from an adult somatic cell, using the process of nuclear transfer has opened a new era of animal cloning. After it, clones of many other animals like cows and buffaloes have been obtained. Even it is possible to obtain clone of any human.

Dolly was cloned by **Ian Wilmut, Keith Campbell** and colleagues at the Roslin Institute in Edinburg, Scotland. She was born on July 5, 1996 and died on February 14, 2003.

Dolly was produced from an udder cell (mammary gland cell) obtain from its mother as follows:

1. First of all udder cell was isolated from a sheep, the nuclear donor (the mother of Dolly). The nucleus of udder cell of sheep has all basic information to develop into a lamb similar to the sheep from which it is obtained.

2. An egg cell was obtained from another sheep (a cytoplasm donor sheep).
3. Nucleus of udder cell was sucked out with the help of a micropipette by appropriate technique.
4. Similarly nucleus of the egg cell (obtained from a cytoplasmic donor sheep) was also sucked out.
5. Then isolated nucleus of the udder cell was introduced into the enucleated egg cell by electrical stimulation.
6. The egg cell whose nucleus has been replaced by the nucleus of the udder cell was cultured in the laboratory. This was cultured in the laboratory for six days during that period it transformed into a six day initial embryonic stage.

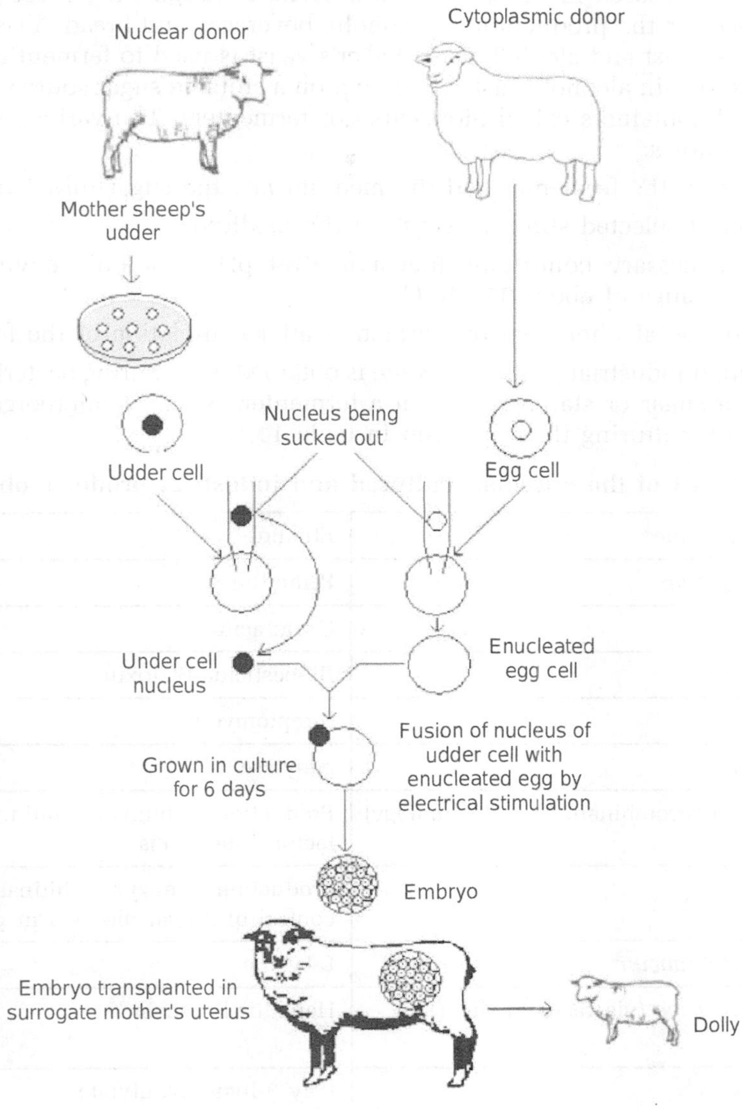

Fig. 11.4. Cloning of Dolly.

7. The six day old embryo was transplanted into the womb of another sheep (surrogate mother). When a normal fertilized egg (zygote) or an altered egg as in the present case is implanted in the womb of another female other then the egg donor. Such females are called *surrogate mother*.
8. Surrogate mother sheep in the appropriate duration of time gave birth to Dolly that was clone of her mother (the sheep from which udder cell was obtained).

Culture of Microbes and Production of Industrial Products

Microbes have been utilized for the production of industrial products of human utility since the advent of human civilization. There is use of both naturally occurring as well as genetically modified microorganisms for the production of desired products. The most common, naturally occurring microorganism used in industries is a unicellular fungus, the yeast (*Saccharomyces cerevisiae*). It is used for the production of alcoholic beverages and bread. Yeasts are basically of two types: baker's yeast and alcohol yeast. Baker's yeast is used to ferment and raise dough in bread making. To obtain alcohol yeasts are grown on a suitable sugar source such as sucrose in suitably designed containers called bioreactors or fermenters. The various steps of alcohol production are as follows:

1. Sterilization of the fermenter and the medium *i.e.*, the sugar dissolved in water.
2. Inoculation of selected strain of yeast on the medium.
3. Providing necessary conditions like a neutral pH, anaerobic environment and a temperature range of about 35^0–40^0C.
4. Recovery of the alcohol from the fermenter after completion of the fermentation.

Another common industrial product acetone is obtained by culturing bacterium *Clostridium acetobutylicum* on a sugar or starch source in a fermenter. A list of microorganisms and the products obtained by culturing them is given in table 11.1.

Table 11.1 List of the microbes cultured and industrial products obtained

Saccharomyces cerevisiae	Ethanol
Penicillium chrysogenum	Penicillin
Aspergillus niger	Citric acid
Bacillus thuringiensis	Biopesticide(Bt toxin)
Streptomyces griseus	Streptomycin
Aspergillus oryzae	Amylase
Escherichia coli (via recombinant DNA technology)	Production of human insulin, human growth factor, interferons
Trichoderma	Production of enzyme chitinase for biological control of fungal diseases in plants
Corynebacterium glutamicum	L-Lysine
Saccharomyces cerevisiae (via recombinant DNA technology)	Hepatitis-B vaccine
Alcaligenes eutrophus	Poly 3-hydroxybutyrate
Propionibacterium shermanii	Vitamin-B_{12}

Exercises

1. What is biotechnology? Give a brief account of its scope and achievements.
2. What is the meaning of totipotency of cell?
3. Define tissue culture.
4. Describe the procedure of plant tissue culture.
5. What are transgenic plants? What are the different methods of gene transfer?
6. What is the meaning of somatic hybridization? Describe the process of obtaining somatic hybrids.
7. What is cloning? What is its significance?
8. How "Dolly" the first mammalian clone was obtained?
9. What is a bioreactor?
10. Write a note on microbial culture.

❑❑❑

Chapter 12 APPLICATIONS OF BIOTECHNOLOGY

Biotechnology is the most exciting revolutionary science of the present day. It has opened up new horizons in the field of biological sciences. It has immense applications in almost every sphere of human life. It has wide spread applications in the field of agriculture, medicine and redressing environmental issues like waste management and pollution prevention.

APPLICATION IN AGRICULTURE

Biotechnology has helped in the improvement of crop plants. Many transgenic crop plants have been obtained which are disease resistant, pest resistant, and able to sustain environmental stresses like resistant towards salinity and drought by biotechnology. All these methods have resulted in increase of food production, increase in the quality of food and decrease in the cost of production of crop plants. Some of the important role of biotechnology in the field of agriculture is as follows:

1. Increase in Quantity and Quality of Plant Yield

Food is the basic requirement of all living organisms including human beings. Demand of food is increasing day by day due to population explosion especially in the third world countries but the area of cultivated land is limited. Food production can not be increased only by the traditional methods of agriculture. Experts assert that using biotechnology in agriculture can triple the food production without requiring any additional farm land. The large scale use of chemical fertilizer, pesticides and fungicides has polluted the land, water and air. Here biotechnology is at rescue to reduce dependence on the use of chemicals used in agriculture and thus reducing the pollution as it encourages organic farming. Many high yielding and disease resistant varieties of crop plants have been obtained by genetic engineering and tissue culture techniques which have helped in many fold increase in the yield of crop plants.

2. Micropropagation (Rapid Multiplication of Plants)

With the help of plant tissue culture technique a very rapid multiplication of desired good quality plants is possible, in this technique a small tissue obtained from a selected plant is cultured in the laboratory on nutrient medium where it give rise to hundreds of somatic embryos, each of which is grown into a plant, and such plants are transferred from the laboratory after hardening into the fields and they grow successfully in the field and are produced at a comparative low cost. This method of multiplying plants by tissue culture technique is known as micropropagation. It is a very good method of plant replication especially in those cases where plants are multiplied by vegetative propagation which is labor intensive and takes more time. By this technique good quality plants like oil palm, banana, cassava, camphor, and mango are multiplied.

3. Synthetic Seeds

Synthetic seeds or artificial seeds are obtained by plant tissue culture. Synthetic seeds are the encapsulated somatic embryos which functionally mimic seeds and can develop into seedling under sterile conditions. In this technique the somatic embryos obtained by tissue culture of selected plants are encapsulated into calcium alginate gel containing essential nutrients and antimicrobial agents. Artificial seeds can be stored or transported under appropriate condition and are used for rapid multiplication of desired plants.

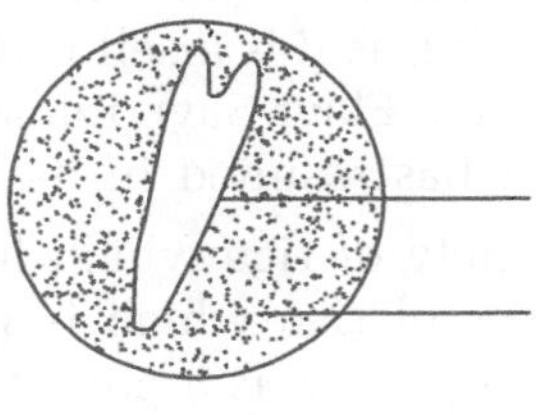

Fig. 12.1. Synthetic seed.

4. Disease Resistant Crop Plants

There are many types of plant diseases caused by viruses, bacteria, fungi and nematodes. Plant diseases cause heavy losses to the food crop plants. Traditional methods of disease control is the use of chemicals like fungicides , bactericides etc. which are loosing their efficacy as the pathogenic organisms develop resistant to the fungicides and other chemicals when they are used continuously to control them. Moreover, the chemicals used to control plant diseases pollute the environment. With the help of genetic engineering disease resistant varieties of crop plants are developed which is a good method of disease control and also reduces the dependence on chemicals for disease control. For example, disease resistant virus plants have been developed which are resistant to viral diseases. To develop a virus resistant crop plant, a gene coding for protein coat of the virus is isolated from its genome and is introduced in the plants with the help of suitable vector. Once the viral gene coding for the protein coat is inside a plant it expresses and in response to it natural resistant against the virus is developed in the plant and the plant becomes resistant to that very virus. Using this technology the first successful virus resistant crop, a virus resistant tobacco was introduced in China in 1990. Similarly other crop plants are made resistant to plant diseases.

5. Transgenic Crops

Many transgenic plants such as pest resistant Bt. Corn and Bt. Cotton; and improved nutrient quality plants like Flavr Savr Tomato and Golden Rice have been developed which are much superior to the traditional crop plants. A brief account of these plants is as follows:

(*i*) **Bt Corn.** Corn plants are attacked by some insects like Corn Bores which cause huge losses to this crop. Corn Bores are difficult to control as they have become resistant to insecticides and moreover they escape from the effect of insecticides as they tunnels into corn stalks after hatching and is protected from insecticide. Bt Corn contains a gene introduced into it from a bacterium *Bacillus thuringiensis* that produces a toxic protein and insects are killed by it. Bt Corn is pest resistant and has decreased the dependence of insecticides to control the insects and has also reduced pollution.

(*ii*) **Bt Cotton.** Bt Cotton is transgenic cotton which is resistant to bollworms in which a gene from the bacterium *Bacillus thuringiensis* has been introduced as in the case of Bt corn. Bollworms causes heavy losses to the cotton crops, the large scale use of insecticides has resulted into resistant of bollworms to the insecticides. Bt Cotton has allowed reducing insecticide use and attaining better control of the insect bollworms.

(*iii*) **Flavr Savr Tomato.** It is a transgenic tomato which has a longer shelf life and has better flavour as compared to the common ones. The traditional tomatoes after ripening can not stay longer as they have an enzyme called *'polygalacturonase'* which is responsible for fruit softening as it is a cell wall degrading enzyme and thus reduces its shelf life. In the transgenic Flavr Savr Tomato the amount of the enzyme is reduced to a low level and that has resulted in its long shelf life.

(*iv*) **Golden Rice.** Golden Rice is a variety of rice which has much more β-carotene as compared to the traditional varities of rice which is pink in colour and has been obtained by genetic engineering. β-carotene is a precursor of vitamin-A, the vitamin for the good eyes. Golden Rice is developed as a fortified food to be used in areas where there is shortage of dietary vitamin-A. Deficiency of vitamin-A is a cause for night blindness and Golden Rice is treated as a boon to solve the problem of night blindness especially in the third world countries.

(*v*) **Hirudin Rich *Brassica napus*.** Hirudin is a chemical which is present in human blood and blood of other animals and it checks the clotting of blood. *Brassica napus* is a common oil yielding crop. A gene coding for hirudin, is isolated from human genome and has been introduced in *B. napus* with the help of a vector (bacterium) to develop a transgenic variety of it , and its seeds are used to extract hirudin which is used to generate blood thinner employed in medicine to dissolve the blood clots.

6. Biofertilizers

Biofertilizers are the biologically active microbial inoculants of such beneficial microorganisms which on their application to seed, or soil help in the availability of nutrients particularly the nitrogenous compounds for the crop plants. Biofertilizers includes nitrogen fixing bacteria, nitrogen fixing blue green algae, *Azolla*, green manure and mycorrhizae.

(*a*) **Nitrogen Fixing Bacteria.** There are two main types of nitrogen fixing bacteria symbiotic (*Rhizobium*) and free living *Azotobacter* (aerobe) and *Clostridium* (anaerobe). The nitrogen fixing ability of these bacteria is employed to increase the nitrogen availability to plants by natural means. For example, *Rhizobium*, which occurs in the soil forms nodulated roots in leguminous crop plants and provides nitrogenous compounds to them. Now more efficient strains of *Rhizobia* are available, they are mixed with the seeds of legume plants before sowing and subsequently these strains of Rhizobia fixes nitrogen more efficiently and as a result of which legumes become more productive.

(*b*) **Nitrogen Fixing Blue Green Algae.** There is a number of nitrogen fixing blue green algae (cyanobacteria) used as a biofertilizer. Paddy crop (rice crop) is grown in a constant standing water and algae occurs in water naturally, the yield of paddy crop is enhanced considerably by the use of nitrogen fixing cyanobacteria like *Nostoc, Anabaena, Spirulina* etc., the inoculants of these cyanobacteria are available in ready to use packets. These cyanobacterial inoculants are mixed in the water of paddy fields. These cyanobacteria grow and multiply there and enrich the soil nitrogen utilized by paddy plants and helps in increasing yield of paddy.

(*c*) **Azolla.** *Azolla*, a water fern, which grows naturally in water bodies like ditches, ponds etc. is a good source of biofertilizer as in its body a nitrogen fixing blue green alga *Anabaena* occurs as a symbiont, when *Azolla* is added into the fields it enriches the nitrogen contents of the soil and helps in increasing the yield of crop plants.

(*d*) **Green Manure.** Green manure is used to increase the fertility of soil. It is obtained from the leguminous plants. Some of the legume plants used for obtaining green manure are *Sesbania* ('dhaincha'), 'guar' (*Cymopsis tetragonoloba*), and ground nut (*Arachis hypogea)*. These plants are grown in the field and when flowering is initiated in them they are ploughed in the field. As these plants have nodulated roots in which *Rhizobium*, a symbiotic bacterium is present that fixes nitrogen as mentioned earlier. The debris of these plants is rich in nitrogenous compounds which are acted upon by saprophytic soil bacteria. The decomposed plant debris is converted into manure referred as green manure that helps in increasing the fertility of soil.

(*e*) **Mycorrhizae.** Mycorrhizae are the fungi which are associated with the roots of higher plants. These fungi help in solubilization of nutrients chiefly the phosphorus and make their availability to the plants. These fungi also provide protection to the plants from disease causing organisms. An important mycorrhizal fungus *Trichoderma* especially protects the plants from pathogenic organisms and efforts are made to provide ready made inoculants of this fungus which can be added in the soil to provide protection to the plants from disease causing organisms.

7. Biopesticides

Biopesticides are obtained from microorganisms and plants and are used to control the menace of pests both in fields and during storage of grains and seeds. A soil bacterium, *Bacillus thuringiensis*, is a very good source of biopesticides. Its spores produce the insecticidal Cry protein that kills larvae of harmful insects including ballworms. Ballworms cause large scale destruction in cotton crop and in severe cases whole of the crop is ruined. Due to the failure of cotton crop, hundreds of farmers committed suicide in Andhra Pradesh in 1999. The commercial preparation of *B. thuringiensis* is consists of a mixture of its spores, Cry protein and an innert carrier. This bacterium is the first biopesticide produced from a microorganism on commercial scale and is also the first such product to be produced at commercial scale in India. Another well known biopesticide is azadirachtin obtained from neem and is isolated from kernel of its seeds. Neem is a traditional biopesticide in India. Its dry leaves are put in clothes chiefly the woolen ones when they are kept at safe place in houses during the off season. It also provides most effective control of major pests of agricultural crop plants as well as protecting stored seeds against pests. It is environment friendly pesticide and very effective and is highly biodegradable. The use of biopesticides is expecting to reduce the dependence chemicals for control of diseases and pests.

APPLICATION IN MEDICINE

Biotechnology is playing a very significant role in the field of medicine. It is applied in the production of many pharmaceutical products like vaccines, antibiotics, hormones and enzymes which are very vital for human health care. Biotechnological aids like diagnostic kits help the physicians to diagnose the disease in a short duration of time so that treatment of the patient can be initiated early and timely. Gene therapy to cure some hereditary disease is now possible, the hereditary disease are otherwise incurable or very difficult to cure.

1. Diagnostic Kits

A Diagnostic Kits is used to detect the occurrence of a disease in a very short duration of time. At present early detection of many types of diseases like malaria, hepatitis and others are possible. Diagnostic kits are also available for the detection of pregnancy, for checking the sensitive allergens etc. Detection of diseases is done generally by immunological methods employing specific antibodies. A diagnostic kit used to detect the presence of pathogenic organism in the blood of a patient is basically consists of a specific monoclonal or polyclonal anybodies raised against the antigen of the pathogenic organism. For example, to detect the presence of malarial parasite in the blood of a patient, his blood is incubated along with the antibodies of malaria, formation of precipitates indicate the presence of malarial parasite. With the help of diagnostic kits early detection of diseases become possible so that treatment can be started in time.

2. Vaccines

Vaccines are the biological preparations which are used to activate immune system of the body to fight against the pathogenic organisms. Vaccines act as antigens, when they are injected in the body, in response to them B-lymphocytes of blood generates antibodies that fight against infectious agents. Vaccines are consist of killed or attenuated (weakened) live microorganisms or their toxins and are either injected in the body or introduced through mouth or by nasal spray. **Edward Jenner** discovered the first vaccine against small pox in 1796. Vaccines have been developed against disease caused by bacteria (*e.g.*, typhoid, whooping cough, tuberculosis) and viruses (*e.g.*, measles, influenza, rabies, and poliomyelitis).

With the help of recombinant DNA technology the efficacy of vaccine can be improved to give better results. As explained above the conventional method of vaccine formation is use of killed or weakened microorganisms (bacteria and viruses). Often it has been observed that some of the components of vaccines have deleterious effects such as fever etc. Since proteins are the main molecules which provides stimulus for immunity attempts have been made to engineer proteins having minimum deleterious effects. The specific sequences of amino acids in proteins which stimulate immune response are known as *epitopes*. With the help of recombinant DNA technology it is possible to synthesize a vaccine, the recombinant vaccine based on selected epitopes which may provide the safety needed for an effective vaccine. Working on such lines a novel synthetic gene has been assembled to synthesize vaccine against hepatitis B virus (HBV) infection.

3. Insulin

Insulin is a hormone which regulates blood sugar in human and it is synthesized in the pancreas, a gland present in human body. Person suffering from diabetes mellitus can not synthesize insulin and they require a regular dose of insulin which is injected in their body to regulate the proper blood sugar level in their body. Traditionally the exogenous source of insulin is to extract it from pancreas of animals like pigs; goats etc. obtained from the slaughter houses. This is not a proper way of obtaining insulin moreover at present its demand cannot be met with the insulin extracted from pancreas of slaughtered animals. Now, with the help of genetic engineering, the insulin gene isolated from a rat or human being is transferred into the plasmid of *E. coli,* a bacterium in which it expresses and synthesizes the insulin. Bacteria multiply at a faster rate so they acts as 'mini insulin factories' and is a very good source of hormone insulin.

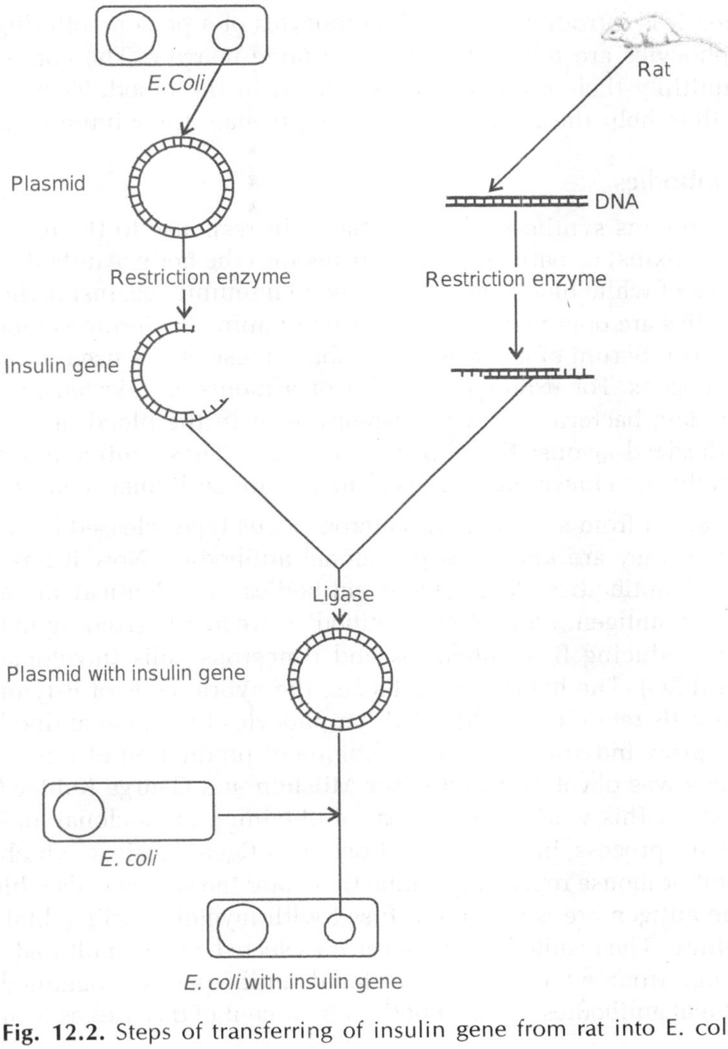

Fig. 12.2. Steps of transferring of insulin gene from rat into E. coli.

4. Gene Therapy

Gene therapy is an upcoming new biotechnological method of treatment in which a defective gene present in human cell is replaced with a functional normal gene mediated through a specifically designed suitable vector (gene carrier). It is still at experimental stages. One of the defective genes that have been successfully replaced is the gene coding for an enzyme *Adenosine deaminase* (ADA). Function of the enzyme ADA is to detoxify toxins and thus helps to combat with the pathogens and provide protection against disease causing agents. It is secreted by B-lymphocytes of blood. In a few people the enzyme ADA is not synthesized in their B-lymphocytes due to the defect in the gene responsible for it. Such person has a weak body immune system and suffers from *Severe Combined Immunodeficiency Syndrome* (SCID) and is easy prey of different pathogenic organisms. It has become possible to correct this defective gene. The lymphocytes of a person suffering from SCID are extracted from his bone marrow. The defective gene is replaced with a functional gene. A functional gene from the B-lymphocytes of normal healthy human being is extracted and is put in the genome of a retrovirus by appropriate techniques, and then it is introduced in the lymphocytes of a person suffering from SCID, and the corrected lymphocytes are reintroduced in the bone marrow. The corrected lymphocytes start replicating, multiply their number and are added in the blood. They start secreting the enzyme ADA and thus help the SCID patients to rejuvenate their immune system.

5. Monoclonal Antibodies

Antibodies are the proteins synthesized in the blood in response to the entry of antigens *i.e.,* a 'foreign' substance (toxins) or pathogenic organisms into the body. Antibodies are synthesized by the B-lymphocytes of white blood cells, and provide immunity against pathogenic organisms. Traditionally antibodies are obtained from the serum of animals. Serum is blood plasma except the blood protein fibrin. Serum of animals like rabbit, mouse etc. is used to immunize humans against specific pathogens. For example, a rabbit or a mouse is injected by an attenuated or weak strain of a cholera bacterium and in response to it in the blood of the rabbit or mouse antibodies are synthesized against the cholera bacterium. These antibodies are isolated from the serum of the rabbit or mouse and are used to immunize human beings against cholera.

Antibodies obtained from serum are of heterogeneous type released by different groups of B-lymphocytes and so they are known as polyclonal antibodies. Now it has become possible to obtain monoclonal antibodies. Monoclonal antibodies are identical molecules which are specific for one type of antigen. Monoclonal antibodies are synthesized by hybridoma cell line by fusing antibody producing B-lymphocytes and cancerous cells (myeloma) in presence of polyethylene glycol (PEG). The hybridoma cells *i.e.,* the hybrid cells of B-lymphocytes and the cancerous myeloma cells retain the ability of B-lymphocytes to secrete antibody and the ability of myeloma cells to grow indefinitely. This technique of production of monoclonal antibodies using hybridoma cells was developed by **Ceasar Milstein** and **George Kohler** (1975) who were awarded Noble Prize for this work. The process of obtaining monoclonal antibodies is shown in the Fig. 12.3. This process involves injection of antigen (against which an antibody is required) into a rabbit or mouse repeatedly. Some time later the spleen cells which are producing antibody against the antigen are isolated and fused with myeloma cells (obtain from cancer of bone marrow) in culture. The resulted hybridomas are selected and are cultured in the bioreactors in the laboratory and from which the monoclonal antibodies are obtained by appropriate techniques. Monoclonal antibodies are used both in treatment of diseases as well as in diagnosing of diseases.

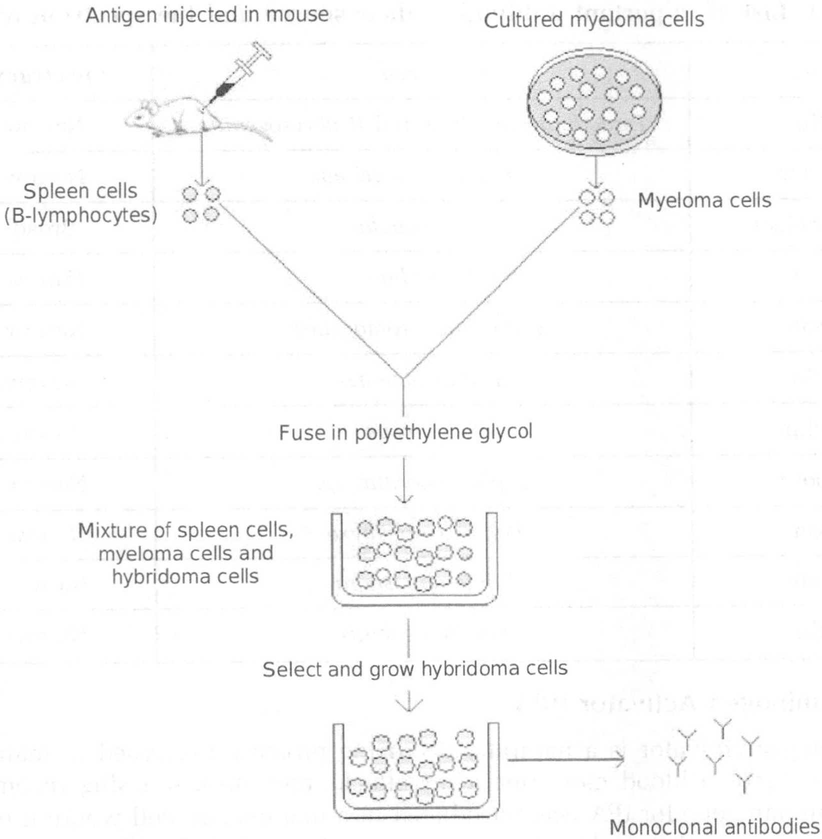

Fig. 12.3. Production of monoclonal antibodies.

6. Antibiotics

Antibiotics are the chemical substances secreted by microorganisms which inhibit the growth of other microorganisms. The penicillin is the first antibiotic obtained from the mold fungus *Penicilium notatum* by **Alexander Fleming** in 1928, since then thousands of different antibiotics have been obtained which are effectively used to control many diseases. Antibiotics are generally of two types: broad spectrum and narrow spectrum antibiotic. The full range of microorganisms attacked by an antibiotic is called its spectrum. A broad spectrum antibiotic is effective against several different types of pathogenic organisms whereas a narrow spectrum antibiotic is effective against some specific families of pathogenic organisms. Tetracyclin and chloramphenicol are the examples of broad spectrum antibiotic and penicillin is an example of narrow spectrum antibiotic. A list of some commonly used antibiotics, the microorganism from which they are obtained and spectrum of their activity is given in table 12.1.

Table 12.1. List of important antibiotics, their source and the spectrum of activity

Antibiotic	*Source*	*Spectrum of activity*
Penicillin	*Penicillium notatum* and *P. chrysogenum*	Narrow spectrum
Streptomycin	*Streptomyces griseus*	Narrow spectrum
Chloromphenicol	*S. venezuelae*	Broad spectrum
Neomycin	*S. fradiae*	Narrow spectrum
Oligomycin	*S. diastatochromogenes*	Narrow spectrum
Linomycin	*S. linocolnensis*	Narrow spectrum
Tetracyclin	*S. rimosus*	Broad spectrum
Cephalosporin	*Cephalosporium sp.*	Narrow spectrum
Polymyxin	*Bacillus polymyxa*	Narrow spectrum
Gentamycin	*Micromonospora*	Narrow spectrum
Bacitracin	*Bacillus subtilis*	Narrow spectrum

7. Tissue Plasminogen Activator (tPA)

Tissue plasminogen activator is a naturally occurring protease expressed in mammalian cells that is used to dissolve blood clot after heart attacks and strokes. Using recombinant DNA technology a human gene for tPA was introduced in a mammalian cell where it expresses and synthesizes the enzyme tissue plasminogen activator. It was the first drug to be produced by mammalian cell culture. With the help of recombinant DNA technology number of substances of therapeutic value are produced by culturing of genetically altered mammalian cells.

8. Human Growth Hormone

Human growth hormone is synthesized naturally in anterior lobe of pituitary gland attached to the ventral surface of the diencephalon of the brain. In some cases due to some abnormality human growth hormone is not synthesized in appropriate amounts as a result of which infants remain dwarf. Now with the help of recombinant DNA technology it is possible to introduce a gene coding for human growth hormone into the bacterium *E. coli*, which are cultured to produce human growth hormone that is used to treat the affected infants.

9. Blood Clotting Factor VIII/IX

Some human beings suffer from a hereditary disease called haemophilia-A/B due to the inability of synthesizing blood clotting factor VIII/IX as a result of which after cut or other injury blood flow does not stop naturally due to the lack of synthesizing of the blood clotting factor. With the help of the recombinant DNA technology the gene coding for factor VIII/IX is introduced in mammalian cells which are cultured to produce the blood clotting factors which are used to treat the person suffering from haemophilia-A/B.

APPLICATION IN ENVIRONMENT

Biotechnology is contributing a significant role in cleaning of effluents and treatment of wastes. It involves the utilization of microbes to protect or restore natural environment from chemical damages due to human activities. Although nature itself is capable of correcting the damages done to the environment by human activities, biotechnology aids in evolving better strategies by experimenting in the laboratory to reduce the level of effluents and pollutants by generating such microbes that can be utilized in an effective manner to redress the threat to the environment due to fast industrial growth and large scale use of fossil fuels. Some important applications of biotechnology in environment are as follows:

1. Biogas

Biogas is a mixture of gases that is obtained by the anaerobic decomposition of biomass. It is an efficient method of disposing of organic wastes and the biogas generated during decomposition is utilized as a fuel. It consists of about 65 percent methane, 34 percent carbon dioxide and traces of other gases like hydrogen, nitrogen and hydrogen sulphide. Biogas is an efficient fuel and its fuel nature is mainly due to the methane gas. Biogas technology is gaining its acceptance as a self sustaining source of energy both in rural and urban areas. It is one of the fast coming alternative sources of energy in agricultural countries like ours. Millions of individual biogas plants have been installed in our country. Biogas is obtained from waste materials like wastes from animals (cattle dung and urine, poultry droppings, slaughter house wastes etc.), wastes from agro-based industries (bagasses, bran, press-mud from sugar manufacturing, wastes from fruit and vegetable processing etc.), human wastes(faeces and urine), urban solid wastes (house hold residue, waste paper, trash etc.) and some aquatic weeds like water hyacinth.

Biogas is obtained from the wastes by digesting them with the help of naturally occurring anaerobic saprophytic bacteria in a biogas plant. A biogas plant is basically consists of an inlet tank, a digester, and an outlet tank all made up of bricks. A digester of a required size is constructed underground and its upper surface is covered by a dome shaped covering usually made up of steel. Inlet and outlet tanks are constructed on the soil surface. In the digester,

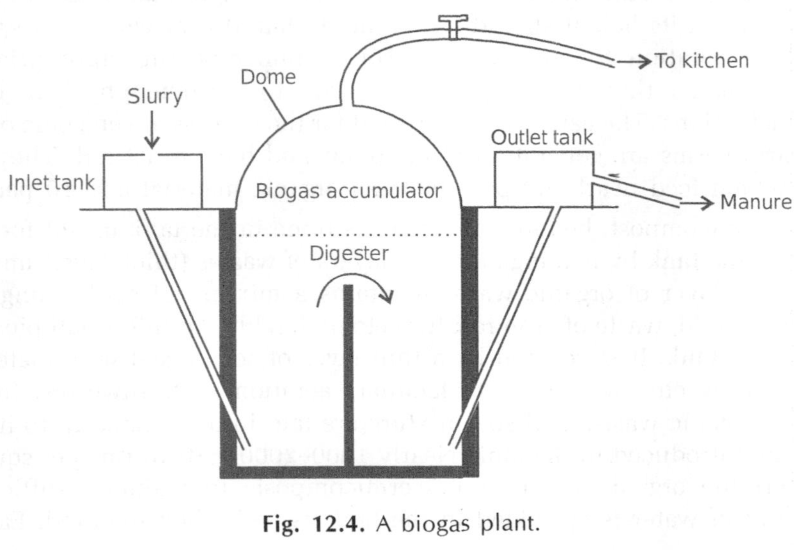

Fig. 12.4. A biogas plant.

slurry (wastes mixed with water) is introduced from inlet tank with the help of a pipe. Function of the digester is to provide oxygen free environment to the anaerobic bacteria for decomposition of wastes. Biogas is generated when organic wastes are decomposed and it takes about 10–15 days for the digestion of the slurry. Biogas starts accumulating in the dome of the digester from where it is taken to a place such as kitchen with the help of a pipe where it is used as a smokeless and hygienic fuel. The digested slurry is passed in the outlet tank through a pipe and that is utilized as mineral rich good quality manure. Fresh slurry is added in the digester at a fixed interval of time.

Anaerobic digestion of wastes to generate biogas is a three stage process completed in a sequential manner by the action of hydrolytic, acetogenic and methanogenic bacteria. In the first stage hydrolytic bacteria like *Bacillus, Clostridium, E. coli* etc. convert carbohydrates (cellulose, hemicelluloses), proteins and lipids present in the wastes into their monomers by the action of hydrolytic enzymes. In the second stage the monomers are converted into organic acids mainly the acetic acid by the action of acetogenic bacteria like *Desulphovibrio, Syntrophomonas* etc. In the last and final stage methanogenic bacteria like *Methanobacterium, Methanobacillus* etc. convert the organic acids into methane gas. During fermentation carbon dioxide and traces of other gases like H_2, N_2 and H_2S are also formed.

2. Vermicomposting

It is a technique of converting organic wastes into manure through the action of earthworms. This technique enables in the quick disposal of organic wastes in which earthworms along with the microorganisms degrade the organic matter into mineral rich compost that is used in increasing the fertility of the soil. It is an environment friendly technique that is gaining fast popularity. An additional advantage of this technique is that the earthworms multiply during the process of vermicomposting are used as a protein rich food supplement for livestock industry.

Vermicomposting is done in a box or a tank made up of bricks or stones. The length, breadth and height of the tank can be according to the convenience and amount of the waste available for vermicomposting. An ideal small sized tank required for vermicomposting is 2m in length, 1.5m in breadth and 0.75m in height. However, length and breadth of the tank can be increased further but its height should not be more than 0.75m which is required in order to facilitate proper aeration for earthworms. For vermicomposting appropriate species of earthworms are selected that have high adaptability to extreme physical, chemical and environmental fluctuations. The earthworms selected for the process are epigenic or detritivorous worms. These earthworms are generally red in colour and have a flat tail. They are called as humus formers which feed nearly 90 percent fresh organic material and 10 percent soil.

To prepare vermicompost, bedding of wastes is done in the tank meant for this purpose. Bedding is done in the tank by making alternate layers of wastes (thick layer) and soil mixture (thin layer). A thick layer of organic wastes contains a mixture of cattle dung, wastes from industries and household, waste of crop residues etc preferably cut into small pieces that is laid on the bottom of the tank. It is covered by a thin layer of soil mixed with materials like saw dust, dry leaves, straw etc., which are to facilitate aeration of earthworms. In this manner alternate layers of organic wastes and soil mixture are laid into the tank up to its brink. Then the earthworms are introduced in the tank. Nearly 1000–2000 earthworms per square meter are enough to convert the organic wastes into vermicompost. To maintain sufficient moisture appropriate amount of water is sprinkled in the tank as and when required. Earthworms are voracious feeders that can consume 10–20 percent biomass of their body weight per day. They lay castings or the faecal matter that is used as good quality mineral rich manure. The process of vermicomposting is completed in about 6 week. After that the compost and the worms are separated manually. The earthworms can be employed time and again for the process of vermicomposting.

3. Bioremediation

Bioremediation means cleaning the environmental pollutants like oil, gas, heavy metals and other toxic substances from the soil or water by the use of living organisms. Some microbes have an appetite for gas, oil, or other toxic chemicals, and such microbes either degrade the pollutants or accumulate in their body by converting them into some non toxic forms and thus help in cleaning the environment and returning it to its natural state. In a sewage disposal plant it is essentially the microbes chiefly the bacteria that degrade the pollutants and other waste materials and finally the sewage is converted into manure that is utilized to increase the fertility of the soil.

To make some microbes more efficient in breaking down the pollutants, genetic engineering is employed to add new characters to the microbes. Working on these lines a genetically altered strain of bacterium, *Pseudomonas putida* was developed by **Anand Mohan Chakroverty,** an Indian born and the US based scientist in 1986 that is effectively used to clean the oil spills from the seas.

Another method of cleaning the pollutants from the soil is by growing plants on the degraded land and this technique of cleaning the environment is known as *phytoremediation*. Specially selected or engineered plants are used in this process. Phytoremediation is an upcoming promising technology that is used in reducing the hazardous heavy metal pollutants present in contaminated soil and surface or ground water by growing such plants which accumulates the heavy metals in their body; and then can be extracted or disposed of safely from the plant body. One of the main advantages of this technique is that it is relatively low cost as compared to conventional practice such as excavation. Some of the plants which are used in phytoremediation are aquatic weeds like *Salvinia, Lemna, Azolla and Eichhornia*, sedges like *Typha* and some herbaceous plants like Indian mustard as well as woody flowering plants like *Poplar* and *Eucalyptus*.

BIOTECHNOLOGY AND ITS PUBLIC PERCEPTION

Biotechnology has lots of application and some of which are described in this chapter. In the present time most of the biotechnological research is concentrated on the generation of genetically modified organisms generally referred as GMOs, by the use of recombinant DNA technology. The acceptability of genetically modified food crops and other products formed by recombinant DNA technology is facing stiff resistance by the public in general and the environmentalist in particular due to the fear such as use of such products may be allergic, and GMOs pose a threat to the biodiversity. Public opinion may sometimes not be based on scientific facts, but it can not be ignored fear based on public perception are none the less very real. Some of the possible adverse effects enumerated by the environmentalist against the genetic engineering and the release of GMOs to the environment are as follows:

1. Creation of new allergens and toxin in the environment by the use of GMOs and the products obtained by them. The cry protein which is secreted by pest resistant Bt cotton to save the cotton from the attack of bollworms may become an allergen for other living organisms including human beings.
2. There is a fear of creation of a 'super weed'. The term *'super weed'* is often used to express concern that introduced gene will jump from crop plants to wild relative crops and that new gene will make the resulting hybrid more vigorous and weedy than either of their parents. Controlling of that very weed may become problematic.
3. There is a fear to wild life and biodiversity. Use of GMOs especially in case of crop plants will reduce the dependence on the prevailing cultivars and in the due course of time if only GMOs are preferred the other cultivars will start disappearing.
4. Introduction of a gene from one species to other species violates the 'integrity of the species'.
5. Biotechnology may pose some unforeseen risk to the environment.
6. Recombinant DNA technology may be misused by some notorious enemies of humanity to generate 'super bugs' which may be used as biological weapons.
7. There are some other ethical issues like cruelty against animals *e.g.*, when animals are used for production of pharmaceutical proteins, they are virtually reduced to the status of a factory.
8. Transfer of human gene into animals and vice versa dilutes the concept of 'humanness'.

Exercises

1. Enumerate the applications of biotechnology in the field of agriculture.
2. What are synthetic seeds?
3. Write a note on the transgenic crop plants.
4. Give a brief account of biofertilizers.
5. Write a note on biopesticides.
6. Name the various application of biotechnology in the field of medicine.
7. What are antibiotics? Give one example each of a broad spectrum and a narrow spectrum antibiotic.
8. What are vaccines? Write a brief note on vaccines.
9. Write a brief note on monoclonal antibodies.
10. What is gene therapy?
11. Write a note on production of insulin employing biotechnology.
12. Give the process of biogas production and its importance.
13. Write a note on vermicomposting.
14. What is bioremediation?
15. Write a brief note on public perception of biotechnology.

❑❑❑

INDEX

A

Abscisic Acid, 72
Acclimatization, 162
Active Absorption, 56
Aflatoxin, 108
Alcoholic Fermentation, 105
Allele, 126
Allopatric Speciation, 123
Ammonotelism, 92
Anaerobic Respiration, 79
Aneuploidy, 122
Animalcules, 96
Anthocyanin Pigmentation, 56
Antibiotics, 109, 177
Antimicrobial Agents, 109
Antiseptics, 109
Apoenzyme, 41
Archaea, 99
Atavism, 125
Autoclave, 161
Auxins, 72

B

Bacteriochlorophyll, 10
Bacteriophages, 103
Bilateral Symmetry, 74
Biofertilizers, 172
Bioinformatics, 153
Biological Scissors, 149
Biopesticides, 173
Bioremediation, 181
Blood Clotting Factor VIII/IX, 178
Blood Vascular System, 86
Bt Cotton, 164

C

c-DNA Library, 151
Candida, 107
Capsule, 9
Carboxylation, 66
Cell Division, 134
Cell Membrane, 9
Cellulose, 19
Central Dogma, 148
Centrosomes, 13
Chemosmotic Hypothesis, 65
Chemosynthetic Bacteria, 60
Chlorosis, 56
Chromosomal Aberrations, 122
Chromosome, 140
Chyme, 77
Circulation of Body Fluid, 84
Clones, 166
Coenzyme, 48
Cofactor, 48
Community, 7
Competitive Inhibition, 47
Conjugated Lipids, 22
Conjugated Proteins, 29

Connecting Link, 118
Continuity of Germ Plasm Theory, 119
Control of Microbes, 108
Cryptogams, 53
Cutaneous Respiration, 80
Cyanobacteria, 101
Cytokinins, 72
Cytoplasm, 9, 10

D

Dark Reaction, 66
Darwin's Theory of Natural Selection, 119
Deletion, 122
Denaturation, 27
Denitrification, 61
Deoxyribose Nucleic Acid, 29
Dermatitis, 38
Diagnostic Kits, 174
Diakinesis, 139
Die Back, 59
Diplotene, 139
Disinfectants, 109
Dominant Trait, 129
Drought Rot, 59
Duplication, 122

E

Ecosystem, 7
Embryological Evidences, 115
Endoplasmic Reticulum, 11
Enzymes, 40
Escherichia Coli, 146
Etiolation, 71
Eukaryotic Cell, 10
Exanthema, 59
External Respiration, 80

F

Feedback Inhibition, 47
Fermentation, 40
Flagella, 9, 11
Flevivirus, 107
Fungi, 102

G

G1 Phase, 134
G2 Phase, 135
Gene Library, 151
Gene Therapy, 175
Genetic Engineering, 149
Genetically Modified, 181
Genome, 152
Genomic Library, 151
Genomics, 152
Genotype, 126
Geological Evidence, 114
Germ Theory, 99
Gibberellins, 72
Glycogen, 20
Glycolipids, 23
Gram-negative Bacteria, 9
Gram-positive Bacteria, 9
Green Manure, 173
Growth Regulators, 71

H

Haemorrhage, 38
Halophiles, 100
Hatch and Slack Pathway, 67
Heart Rot, 59
Herbivores, 75
Hershey-Chase Experiment, 145
Hierarchy of Organisation, 6
Holoenzyme, 41
Homeostasis, 15
Homologous Organs, 116
Homozygous, 126
Host Cell Machinery, 146
Human Genome Project, 152
Human Growth Hormone, 178
Human Immunodeficiency Virus, 107
Human Respiratory System, 81

I

Incubation, 162
Induced Fit Hypothesis, 43

Inhibitors, 47
Inoculation, 162
Insulin, 174
Internal Respiration, 80
Interspecific Competition, 120
Inversion, 122

L

Lactic Acid Fermentation, 105
Laminar Flow Chamber, 161
Law of Dominance, 129
Law of Independent Assortment, 131
Law of Segregation, 130
Leghaemoglobin, 60
Leptotene, 139
Ligases, 45
Light Reaction, 62
Little Leaf, 58
Living Organisms, 4
Lizard Bird, 115
Lock and Key Hypothesis, 42
Lymphatic System, 91

M

Macronutrients, 54
Mechanism of Respiration, 83
Meiosis, 137
Meristems, 71
Mesosome, 9
Microbiology, 96
Microbodies, 11
Microinjection, 165
Micronutrients, 54
Micropropagation, 162, 171
Mitosis, 135
Monoclonal Antibodies, 176
Monohybrid Cross, 128
Monosaccharides, 17
Mutations, 121
Mycoplasma, 101
Mycorrhiza, 104, 173

N

Necrosis, 56
Nephron, 93
Night Blindness, 37
Non Competitive Inhibition, 47
Nucleoid, 9
Nucleoside, 31
Nucleotide, 31
Nutrient Medium, 160

O

Oesophagus, 77
Oligosaccharide, 19
Open System, 14
Optimal Temperature, 45
Organic Evolution, 113
Organic Farming, 159
Organism, 6
Origin of Life, 3
Osmoregulation, 92
Osteomalacia, 38
Oxidoreductases, 44

P

Pachytene, 139
Particle Gun, 166
Passive Absorption, 56
Penicillin, 177
Peptidoglycan, 8
Phanerogams, 53
Phosphodiester Bond, 33
Phospholipids, 22
Photophosphorylation, 65
Phytoremediation, 181
Pleura, 82
Polypeptide, 27
Polyploidy, 123
Polysaccharides, 19
Prophase, 135
Proteins, 23
Protoplasm, 4
Psychrophiles, 100

Pulmonary Respiration, 80
Purines, 31
Pyrimidines, 31

R

R-type, 144
Radial Symmetry, 74
Recessive, 126
Recessive Trait, 129
Recombinant DNA Technology, 149
Recombination, 122
Reductive Ammination, 61
Regeneration, 67
Repair and Regeneration, 6
Ribose Nucleic Acid, 29
Ribosomal RNA (rRNA), 35
Ribosomes, 9

S

S-Phase, 135
S-type, 144
Saprophytic, 59
Sedoheptulose, 18
Seed Germination, 70
Serum, 88
Simple Proteins, 28
Sinigrin, 58
Solarization, 69
Somatic Hybridization, 162
Speciation, 123
Steady State, 14
Sterility, 38
Sticky Ends, 149
Stomach, 77
Sucrose, 19
Super Weed, 182
Surrogate Mother, 168
Symbiotic, 59
Sympatric Speciation, 123
Synthetic Seeds, 171

T

Telophase, 137
Thermoacidophiles, 100
Thermo Labile, 109
Tissue Plasminogen Activator, 178
Tobacco Mosaic Virus, 107
Totipotency, 160
Trans Ammination, 61
Transcription, 35
Transfer RNA (tRNA), 35
Transferases, 44
Transforming Principle, 145
Transgenic Crop, 159
Transgenic Plants, 164
Translocation, 122
Turgidity, 71
Turnover Number, 41

U

Ureotelism, 92
Uricotelism, 93
Urinary Bladder, 94

V

Vaccines, 174
Variation, 120, 121
Vector, 149
Vector Mediated Gene Transfer, 164
Vermicomposting, 180
Vestigial Organ, 117
Viruses, 103

X

Xanthomonas Citri, 107
Xerophthalmia, 37

Z

Zygotene, 139

❑❑❑